# Conversations in Human Evolution
## Volume 2

# About Access Archaeology

*Access Archaeology* offers a different publishing model for specialist academic material that might traditionally prove commercially unviable, perhaps due to its sheer extent or volume of colour content, or simply due to its relatively niche field of interest. This could apply, for example, to a PhD dissertation or a catalogue of archaeological data.

All *Access Archaeology* publications are available as a free-to-download pdf eBook and in print format. The free pdf download model supports dissemination in areas of the world where budgets are more severely limited, and also allows individual academics from all over the world the opportunity to access the material privately, rather than relying solely on their university or public library. Print copies, nevertheless, remain available to individuals and institutions who need or prefer them.

The material is refereed and/or peer reviewed. Copy-editing takes place prior to submission of the work for publication and is the responsibility of the author. Academics who are able to supply print-ready material are not charged any fee to publish (including making the material available as a free-to-download pdf). In some instances the material is type-set in-house and in these cases a small charge is passed on for layout work.

Our principal effort goes into promoting the material, both the free-to-download pdf and print edition, where *Access Archaeology* books get the same level of attention as all of our publications which are marketed through e-alerts, print catalogues, displays at academic conferences, and are supported by professional distribution worldwide.

The free pdf download allows for greater dissemination of academic work than traditional print models could ever hope to support. It is common for a free-to-download pdf to be downloaded hundreds or sometimes thousands of times when it first appears on our website. Print sales of such specialist material would take years to match this figure, if indeed they ever would.

This model may well evolve over time, but its ambition will always remain to publish archaeological material that would prove commercially unviable in traditional publishing models, without passing the expense on to the academic (author or reader).

# Conversations in Human Evolution

## Volume 2

Edited by
**Lucy Timbrell**

Access Archaeology

ARCHAEOPRESS PUBLISHING LTD
Summertown Pavilion
18-24 Middle Way
Summertown
Oxford OX2 7LG
www.archaeopress.com

ISBN 978-1-78969-947-0
ISBN 978-1-78969-948-7 (e-Pdf)

# Contents

## Part 4: Evolutionary anthropology and primatology



## Part 5: Human-disease evolution

# Conversations in Human Evolution

Lucy Timbrell[1]

## Introduction:

Conversations in Human Evolution is a science communication initiative exploring the breadth and interdisciplinarity of human evolution research at a global scale. Through informal but informative interviews (henceforth referred to as 'conversations'), this project delves deeply into topics concerning the study of our species' evolutionary lineage, covering the current advances in research, theory and methods as well as the socio-political issues rife within academia. This project also provides important insights into the history of human evolutionary studies, and how methods and ideas have changed over time. Conversations in Human Evolution Volume 1 (Timbrell, 2020) was published last year and reported the first twenty conversations from the project. This volume is the result of the second twenty conversations from the project, published online between June and December 2020. When this volume went to press in March 2021, this subset of conversations had been collectively viewed 6846 times since they were made available on the website. Overall, the website has had 19477 views from 119 countries, with the majority of viewers from the United States of America, the United Kingdom and India.

The idea for Conversations in Human Evolution (CHE) arose in March 2020 during the escalation of the COVID-19 global pandemic. Following the cancellation and postponement of in-person events, CHE became a creative project to encourage engagement with human evolutionary research during this time of isolation and confinement. It was noticed that, whilst there is great public interest in this area of research, there are few freely accessible online resources about human evolutionary studies itself (though see https://humanorigins.si.edu/ for a good example of a publicly available resource). What's more, science engagement initiatives are almost always concerned with communicating exciting results and discoveries, and whilst this is obviously the most important aspect of science communication, it often leads to the neglection of the personal experiences of the scholars behind the science. Broader socio-political issues within subject-specific academic circles are also rarely discussed through publicly accessible communicative forums, somewhat depersonalising the science and perhaps even romanticising academia in certain ways. CHE fills this void by asking: What does it actually mean to study and research human evolution in the 21st century?

Human evolution studies, by definition, is a discipline concerned with the deep past. We explore the most pertinent questions about the evolution of humanity, such as those concerning the emergence of complex language and culture. The exploration of such issues allows researchers to look back into our lineage's evolutionary history to better understand our present and our future. Yet, we rarely consider the role of history and personal experience in the shaping of human evolution research. Acknowledging that the history of our discipline and its historical figures deserve focus in their own right is a fundamental premise of CHE as, in the same way that human evolutionary research drives our understanding of our past, present and future selves, historical and personal contexts have driven modern approaches to the deep past. CHE bridges the gap between the research and the researcher, contextualising modern science with personal experience and historical reflection.

[1] Department of Archaeology, Classics and Egyptology, University of Liverpool, U.K.; lucyjt@liverpool.ac.uk

***Themes:***

The twenty conversations featured in this volume can be organised into five non-mutually-exclusive categories based broadly on research area: (1) Palaeolithic archaeology, (2) palaeoanthropology and biological anthropology, 3) earth science and paleoclimatic change (4) evolutionary anthropology and primatology, and (5) human-disease co-evolution. CHE features scholars at various different stages in their careers and from all over the world; in this volume alone, researchers are based at institutions in eleven different countries (namely Iran, India, the United Kingdom, Greece, Australia, South Africa, the United States of America, the Netherlands, Germany, France and Israel), covering five continents.

Section 1 features five conversations with Palaeolithic archaeologists from around the globe, highlighting just some of the research that is furthering our understand prehistoric human behaviour at a huge geographic scale. First, Dr Sonia Shidrang discusses her research aiming to understand patterns of human behaviour in Late Pleistocene settlements of the Zagros Mountains. She also reflects on some of the cross-cultural differences she experienced working within the different worlds of the East and the West. Next, Professor Ravi Korisettar outlines some of his research into the Palaeolithic of India, while also recounting his somewhat rocky path through academia. Moving into European prehistory, Dr Jennifer French describes her latest project exploring the demography and social lives of early hominins, which has resulted in a monograph comprising the first comprehensive synthesis of the population history of Palaeolithic Europe. Next, archaeologist Dr Rebecca Wragg-Sykes promotes her popular science book 'Kindred: Neanderthal Life Love Death and Art', talking us through her writing process. She also describes other initiatives directed at endorsing women in archaeology. Lastly, Professor Nena Galanidou describes her research into the Palaeolithic of southeast Europe, as well as what inspired her passion for human evolution.

*'Human evolution is a book that is continuously being re-written. New finds and new readings of old finds shed new, often unexpected, direct or oblique light on the old threefold question: who are we – where do we come from – how did we get here'* – Professor Nena Galanidou

In the second section, there are five conversations with palaeoanthropologists and biological anthropologists devoted to understanding human evolution and diversity through fossil and skeletal remains. Professor Rainer Grün, a world-renowned geochronologist, starts off this section with an overview of his life's work; this has been dedicated to understanding the chronology of human evolution, and has involved dating some of the most famous hominin fossils. Then, Dr Briana Pobiner describes her research into the evolution of the human diet, concluding with a discussion on her desire to travel back in a time machine and see what was on the palaeo-menu. She also outlines her role as Education and Outreach lead for the Smithsonian Human Origins Program. Next Dr Mirriam Tawane, the first black female South African to obtain a PhD in palaeoanthropology, describes her curatorial role at the DITSONG: National Museum of Natural History, as well as some of her ongoing paleoanthropological and outreach projects in South Africa. After this, Dr Trish Biers also talks us through her work as a curator of human remains, giving some interesting insights into the ethical issues surrounding curating the dead. Human evolutionary biologist, Professor Tanya Smith, next reviews her research into primate teeth. She also describes her book 'Tales Teeth Tell' and some of the initiatives she's involved with to promote women in biological anthropology. Advocating equality is also a strong theme in the final conversation with Professor Rebecca Ackermann. She explains why we need to denounce colonialist attitudes in human evolution research and discusses some of her research into the evolutionary processes that have shaped human phenotypic variation.

*'The story of human evolution has classically been told by men nearing the end of their careers. Yet biological anthropology is so much more than just a boy's club!'* - Professor Tanya Smith

Three conversations with scientists focussed on understanding human evolution in relation to climate change are outlined in Section 3. First, Professor Rick Potts discusses his role in understanding the eastern African archaeological record, particularly via his field site at Olorgesailie in Kenya. Rick describes the development of his Variability Selection Hypothesis – a theory for understanding how humans became more tolerant to environmental change over time. A conversation with Professor Mark Maslin, who devised a similar hypothesis deemed the Pulsed Climatic Variability hypothesis, follows. Mark reflects on his experience as a PhD student and how these have encouraged him to direct and empower students of the London NERC Doctoral Training Partnership. Finally, Dr Yoshi Maezumi describes her ongoing palaeoecological research into understanding how humans interacted with fire in the deep past, and how she advises students to network in order to open as many doors as possible.

*'In a time-machine, I'd prefer to head at least 100 years into the future and be amazed by what the students and colleagues I'll never meet will have discovered and learned about our species' ancestry'* – Professor Rick Potts.

Section 4 includes conversations with researchers working within the fields of evolutionary anthropology and primatology. First, evolutionary anthropologist Dr Duncan Stibbard-Hawkes discusses his work with the Hadza, which is aimed at understanding food-sharing behaviours in hunter-gatherers. He asks why evolutionary anthropology isn't on the national curriculum despite its applicability to everyday life. Moving over to primatology, Dr Ammie Kalan describes her research on great ape behavioural ecology. She also discusses some of the remote methods she is helping to improve, such as passive acoustic monitoring and camera-trapping. Finally, primatologist Professor Lynne Isbell outlines her work on understanding animal behaviour, as well as her Snake Detection theory for explaining the evolution of primate orbital convergence, visual specialization, and brain expansion.

'*Although we like to see ourselves as exceptional, humans are as much the products of evolution as any other species. And while there are many valid frameworks with which to view ourselves, no account of our actions, our minds and our forms is wholly complete without recourse to evolutionary logic*' – Dr Duncan Stibbard Hawkes

The final section focuses on human-disease co-evolution, a topical focus given the current global situation with the COVID19 pandemic. Dr Hila May starts Section 5 by describing some of her work that attempts to understand pathologies in human remains. She outlines how investigating the evolutionary causes of diseases helps us find new ways of preventing them. Similarly, Dr Simon Underdown is interested in understanding how patterns of human-disease interaction in the past can be used to reconstruct human evolutionary patterns and processes. In the penultimate conversation, Simon also discusses some of the most revolutionary technological advancements in the study of human evolution, such as the analysis of ancient biomolecules and 3D models of human remains. Lastly, population geneticist Professor Lluis Quintana-Murci describes some of his current research in understanding how pathogens and infectious diseases have shaped human evolution. He also discusses some of his work on the Out of Africa dispersal and the demographic and adaptive history of the South Pacific.

'*Understanding how natural selection imposed by pathogens has affected the diversity of our genomes is an alternative way to identify genes and biological functions that have play a key role in human survival against deadly infectious diseases, which highlights the value of dissecting the most natural experiment ever done: that of Nature*' – Professor Lluis Quintana-Murci

*Acknowledgements:*

I would like to first thank Dr Matt Grove and Dr Kimberly Plomp for reading initial drafts of this volume. I would also like to express my gratitude to all of those who have contributed to Conversations in Human Evolution and have supported the project so far. I would also like to acknowledge the Arts and Humanities Research Council and The Leakey Foundation for their ongoing support for my work. Finally, I would like to express my gratitude to my sister, Holly Timbrell, who kindly designed the Conversations in Human Evolution logo.

*References:*

Timbrell, L. (2020). Conversations in Human Evolution: Volume 1. Oxford: Archaeopress. Printed ISBN 9781789695854. Epublication ISBN 9781789695861.

# Part 1: Palaeolithic Archaeology

## Dr Sonia Shidrang[2]

Dr Sonia Shidrang is an archaeologist in the Palaeolithic Department of the National Museum of Iran. Sonia has led several field projects in the Central Western Zagros and recently has initiated a fieldwork project in the Southern Zagros to compare the Middle and Upper Palaeolithic sequences in different regions of this Iranian mountain range. Back in 2001, she began her Palaeolithic career as a research assistant at the newly established Centre for Palaeolithic Research of National Museum of Iran and some years later moved to Europe to complete her postgraduate studies in the field of Palaeolithic archaeology. She finished her PhD at PACEA, Bordeaux in 2015 which focussed on the technological discontinuity between the Middle and Upper Palaeolithic of Zagros and the importance of the site formation process in the issues of the Middle Palaeolithic-Upper Palaeolithic transition in this region. She also proposed a new hypothesis suggesting the possibility of short-time coexistence of the Neanderthals and anatomically modern humans in the region, as another potential explanation for the presence of Middle Palaeolithic tools in the beginning of Upper Palaeolithic occupations in the Zagros.

### *What are your research interests and particular area of expertise?*

Broadly speaking, I am interested in understanding human bio-cultural evolutionary processes during the Pleistocene and the ways that different disciplines, like archaeology, palaeoanthropology, palaeogenetics, palaeoecology, etc., can be used together to try to explain these processes. I enthusiastically follow the ever-changing image that emerges from combining the results of all of these studies in different geographical regions.

As a Palaeolithic archaeologist, I am particularly interested in tracing the earliest emergence of Upper Palaeolithic cultures in Iran which in part helps us to understand the dispersal routes of modern humans through the crossroad region of Southwest Asia. I'm trying to find reliable evidence for possible coexistence and contacts of these modern newcomers with Neanderthals at the Iranian sites, particularly in Central and Southern Zagros. The main focus of my research has been on the techno-typological and taphonomic studies of lithic artefacts, the most frequent archaeological material left behind by Pleistocene human populations. I am mainly interested in the reconstruction of the operational sequences of Early Upper Palaeolithic lithic industries as they reflect changes in techno-economic behavioural patterns of modern populations when compared to the late Middle Palaeolithic lithic industries of Neanderthals or perhaps early modern humans in Zagros.

I think it is absolutely fascinating to trace and understand these variable patterns of human interactions with their environment through time and space with the help of different disciplines.

---

[2] Palaeolithic Department of the National Museum of Iran; sshidrang@gmail.com

### *What originally drew you towards Palaeolithic archaeology?*

Nowadays, the love for archaeology grows from very early ages, as kids watch many archaeological science-fiction movies and documentaries as well as reading fantastic archaeology books specifically designed for them. In my case, my love for archaeology started from the early ages as well but in quite different circumstances. I read my first archaeology-related books, which were extremely rare in the chaotic times of post Iran-Iraq war, when I was ten years old. Reading those few books, and later watching movies about Egyptian dynasties, fascinated me and consequently led me to apply for a BA in archaeology at the age of seventeen.

*The Mar Tarik cave excavation, a Middle Paleolithic site in Bisotun, Zagros, Iran (2004).*

Back in 2001 and during my MA, I started to work at two newly founded centres for Achaemenid studies and Palaeolithic research at the National Museum of Iran as a research assistant, using my computer skills that were quite rare at the time. This was the beginning of my lifelong interest in Palaeolithic archaeology and here I focused on learning how to study lithic artefacts under the direction of Fereidoun Biglari, the first Iranian scholar specialized in the field of Palaeolithic archaeology in Iran. These studies intertwined with reading the pioneering works of Deborah Olszewski and Harrold Dibble on the Warwasi lithic assemblages and works of several other pioneering researchers, such as Jean-Guillaume Bordes, introduced me to the topic of the Middle to Upper Palaeolithic transitions and the Aurignacian as the first widespread culture made by modern humans in the vast area of Western Eurasia. The second wave of inspiration that ensured me of my lifelong career in the field of Palaeolithic archaeology came from a trip to south-western France in 2003 where I met several outstanding French prehistorians. Among them were two prominent women, Laurence Bourguignon and some years later Liliane Meignen, who inspired me to choose this path seriously as a female researcher in a male-dominated field, especially in Iran. In 2004, during our joint Iranian-French Palaeolithic project under the direction of Jacques Jaubert and Fereidoun Biglari in Central Zagros, I benefited significantly from working with several prominent Palaeolithic archaeologists.

By 2005, I moved to Europe as an Erasmus Mundus Masters student in Quaternary and Prehistory and some years later enrolled in a PhD at PACEA, the University of Bordeaux, as a Wenner-Gren grantee and started my PhD under the supervision of Jacques Jaubert and Jean-Guillaume Bordes. Alongside working on my PhD project at Bordeaux University, I went back to Iran constantly to work on several projects with my colleagues Fereidoun Biglari (a Palaeolithic archaeologist) and Marjan Mashkour (a zooarchaeologist), two outstanding researchers that I have had the privilege of working with from the beginning of my career.

*Sonia's first inspirations from Paleolithic archaeology of France (Périgord, 2003).*

*Sonia presenting her work during Erasmus Mundus Masters in Quaternary and Prehistory, Ferrara University and Institute for Human Paleontology, Paris (2006).*

I found that it was sometimes difficult to constantly balance life and work within the two completely different worlds of the East and West, but I think dealing with the cross-cultural differences have considerably expanded my perspectives on life as well as my own professional development as an archaeologist who is trying to understand the behavioural patterns of our ancestors in early prehistory.

### *What was your PhD topic? How did you find your PhD experience?*

My PhD topic was on the Early Upper Palaeolithic of Zagros, involving the techno-typological assessment of three Baradostian lithic assemblages from Khar Cave, Yafteh Cave and Pa-Sangar Rockshelter in the Central Western Zagros, Iran. I had the opportunity and privilege to work with two prominent French Prehistorians with deep knowledge of Middle Palaeolithic and Upper Palaeolithic lithic technologies, Jacques Jaubert (Professor of Prehistory) and Jean-Guillaume Bordes (Director of Research) who became my supervisors at PACEA. In my PhD, I mainly focused on tracing the techno-typological changes of lithic artefacts throughout the whole sequence of the Baradostian by analysing the three mentioned lithic assemblages from the west of Iran. I tried to contextualize each lithic industry and detect their techno-typological characteristics and cultural changes synchronically and diachronically which all led to me describing three clear phases for the Baradostian in the study region, of which some of their characteristics were highlighted by the previous work.

I also tried to address these issues from a research historical perspective. I took a closer look at all published reports from Dorothy Garrod's time until the last decade to examine the lithic-based dominant hypothesis of Middle Palaeolithic-Upper Palaeolithic (MP-UP) continuity in all of the excavated sites in Zagros. When all of the chrono-cultural information and their correlations to the stratigraphy were put together, it seemed that despite the dominant hypothesis of MP-UP lithic industrial continuity, the evidence for technological continuity between the Middle and Upper Palaeolithic elements is very scarce. Back in 2012-2013 when I was gaining more insights into the MP-UP sequences of Zagros, increasing evidence for interbreeding between archaic and modern humans during Middle to Upper Palaeolithic transition was revolutionizing our understanding of interactions between different human populations, initially thanks to the ground-breaking advances in Neanderthal genome sequencing. This was the time that I started to look at my collected information on the peculiar mixture of MP-UP elements in Zagros from a different perspective. Besides emphasizing the importance of site formation processes and the mechanical mixing of archaeological remains, I launched another possible hypothesis for explaining the mixture. The hypothesis, as simple as it was, gave us another option to consider: the cultural indicators of two different human populations, Neanderthals and Anatomically Modern Humans, who might have occupied the landscape in a short overlapping timespan. Just recently, I noticed other researchers working in the region started to adopt the hypothesis, some referring to my work and some competing ex-colleagues who preferred to eliminate the source of the idea...

### *After your PhD, what positions have you held and where?*

Prior to my PhD, I worked as a researcher for the Palaeolithic department of the National Museum of Iran, and during my PhD project in France, I remained an associated researcher to this department. At the same time, I also worked as a project archaeologist, leading several Palaeolithic investigations (some of which were salvage projects) in western Iran. After finishing my PhD, I went to Austria for a short-term post-doc at OREA and then came back to Iran to join a newly founded research institute called the Saeedi Institute of advanced studies at Kashan University on a 2-year contract. For now, I am at the National Museum of Iran as a project archaeologist leading and trying to find budgets for my own Palaeolithic archaeological excavation projects, focused on the issues of MP-UP transition and the emergence and development of Early Upper Palaeolithic cultures in the Zagros. I also teach Masters courses related to Palaeolithic archaeology and human evolution at Tehran's Shahid Beheshti University. As well as this, I have published a book on Upper Palaeolithic of Iran in Persian to help Iranian students of prehistoric archaeology understand the objectives of this field of research.

### *What current projects are you working on? Where do you hope these go in the future?*

Over the past several years, we have conducted several Palaeolithic excavations in the Western and the Southern Zagros with my colleagues, Fereidoun Biglari (Director of Palaeolithic Department, National Museum of Iran) and Marjan Mashkour (CNRS, Director of research) and other international colleagues from around the world. We are currently working on the recovered archaeological materials from these excavations and their analysis. One of the cases that we are focused on right now is the results of an extensive salvage project in the Hawraman region in Kermanshah and Kurdistan, which resulted in the discovery and documentation of a considerable number of Palaeolithic sites and significant Middle and Upper Palaeolithic archaeological materials (Biglari and Shidrang, 2019). We hope that the outcomes of these excavations will improve our understanding of the Middle and Upper Palaeolithic settlements of this region and the crucial shift events that led to the major bio-cultural change in human populations during MIS4 and MIS 3.

*The Kenacheh cave excavation, Hawraman, Kurdistan, Iran.*

### *What project or publication or discovery are you most proud of?*

Apart from our ongoing discoveries in the new sites of the western and southern Zagros and our 2005-2008 discoveries in Yafteh cave, I'm glad that I have collaborated in several publications that shed light on our understanding of the major culture of Upper Palaeolithic in Zagros. I am also pleased about my idea on the possibility of Neanderthals and Modern Humans coexistence for a relatively short period of time in Zagros based on the remaining of their cultural indictors, and I'm eager to see whether the improvement of our dating methods and new results will confirm it or not. On the other hand, systematic careful examination of site formation process is inevitable in this regard.

Since my return to Iran, I have been busy with fieldwork and teaching in the past few years and consequently have had less time for finishing all the publications of the results, but a brief overview of my PhD was published in the Springer book series: Replacement of Neanderthals by Modern Human in 2017 (Shidrang, 2018), along with a few other papers (Shidrang *et al.*, 2016; 2020; Shidrang 2009; 2014; Bordes and Shidrang 2009; Otte *et al.*, 2007; 2011).

### *What is your favourite memory from your career?*

The field of archaeology, particularly in our case with Palaeolithic archaeology, is usually full of amazing adventures, discoveries, and pure moments that only can be found in remote and inaccessible areas where we work during our field projects. Apart from all of the excitements and wonderful moments that I have had experienced at the time of important archaeological discoveries, a non-science-related memory comes to my mind right now.

In 2009, during our Palaeolithic surveys in Kermanshah province, I had borrowed my friend's car as our survey's vehicle and took over the driving myself to prevent any damage to the borrowed car. During the surveys, we visited several villages where respecting the strict local traditions and fundamental values, particularly for women, is one of the most important rules in the daily life of inhabitants. Driving that car (totally common in the Iranian cities) with my archaeological outfit and male colleagues in the car was the most unusual event that attracted much attention and curiosity everywhere on our way. I will never forget the astonished and surprised looks on people's faces when they saw such an unusual woman and, to our surprise, their warm and respectful reactions during our conversations with them.

### *What do you think has been the most revolutionary discovery in Palaeolithic archaeology over the last 5 years?*

Undoubtedly, many exciting discoveries have improved our knowledge and understanding of Palaeolithic periods during last 5 years, from the Lomekwi stone tools dating to 3.3 million years ago that considerably drawback the timespan of Palaeolithic archaeology to the many Neanderthals footprints at Le Rozrl of Normandy, France, that enables us to trace the size and composition of Neanderthal groups. But speaking about a revolution in our understanding of Palaeolithic times and societies, I definitely think the sequencing of the Neanderthal genome was a real revolution in human evolution studies and related disciplines in the past decade (even more, considering the work of Richard Green and colleagues in 2008 (Green et al. 2008) that had its root in the related pioneering research of the late 1990s). All these fossil genome sequencings still continue to surprise us each year with new game-changing findings.

But in the geographical area that I work, Iran, I think the most exciting discovery was the first direct evidence for the presence of Neanderthals that came from a small cave called Wezmeh near Kermanshah at the west of Iran. A premolar tooth that belonged to a Neanderthal child between 6–10 years old was found with large numbers of animal fossil remains (Mashkour *et al.*, 2008; Zanolli *et al.* 2019). It is suggested that the child most probably was killed by a carnivore or his carcass was found by a carnivore in the area and brought to the cave. We re-excavated this interesting cave last year and yielded interesting results.

### *What is the best thing about your job and what is one thing you would change if you could?*

The best thing about my job, I think, is the feeling of freedom, discovery, and adventure from our research. What has kept me in this profession, despite all of its difficulties, is the sense of freedom that it gives me to explore the inaccessible areas of the world, and taking the risks associated with this, in an attempt to try to understand past and present societies and correlate them to their environments.

Regarding the question of what I would change if I could.... it certainly would be the ever-growing unhealthy competition and the extreme 'publish or perish' culture in academia that I think is against the integrity of research and its original purpose. The bibliometric process forces and ultimately leads researchers to produce publishable results at all costs.

*Wezmeh cave excavation, Kermanshah, west of Iran (2019).*

***References:***

Biglari, F. and Shidrang, S. (2019). Rescuing the Paleolithic Heritage of Hawraman, Kurdistan, Iranian Zagros. *Near Eastern Archaeology* 82(4). DOI: 10.1086/706536

Bordes, J. G. and Shidrang, S. (2009). La Sequence Baradostienne de Yafteh (Khorramabad, Lorestaan, Iran), In: M. Otte, F. Biglari, and J. Jaubert (eds), *Iran Palaeolithic*. pp. 85-100, Proceedings of the XV World Congress UISPP, Lisbonne, Vol. 28, BAR International Series.

Green, R., Malaspinas, A.S., Krause, J., Briggs, A.W., Johnson, P.L.F., Uhler, C., Meyer, M., Good, J.M., Maricic, T., Stenzai, U., Prüfer, K., Siebauer, M., Burbano, H.A., Ronan, M., Rothberg, J.M., Egholm, M., Rudan, P., Brajković, D., Kučan, Z., Gušić, I., Wilkström, M., Laakkonen, L., Kelso, J., Slatkin, M. and Pääbo, S. (2008). A complete Neanderthal mitochondrial genome sequence determined by high-throughput sequencing. *Cell* 134(3), 416-426. DOI: 10.1015/j.cell.2008.06.021

Mashkour, M., Monchot, H., Trinkaus, E., Reyss, J. L., Biglari, F., Bailon, S., Heydari, S. and Abdi, K. (2009). Carnivores and their prey in the Wezmeh Cave (Kermanshah, Iran): a Late Pleistocene refuge in the Zagros. *International Journal of Osteoarchaeology* 19, 678-694. DOI: 10.1002/oa.997

Otte, M., Biglari, F., Flas, D., Shidrang, S., Zwyns, N., Mashkour, M., Naderi, R., Mohaseb A., Hashemi, N., Darvish, J. and Radu, V. (2007). The Aurignacian in the Zagros region: new research at Yafteh Cave, Lorestan, Iran. *Antiquity* 81, 82-96. DOI: 10.1017/S0003598X00094850

Otte, M., Shidrang, S., Zwyns, N. and Flas, D. (2011). New radiocarbon dates for the Zagros Aurignacian from Yafteh cave, Iran, *Journal of Human Evolution*, 61, 340-346. DOI: 10.1016/j.jhevol.2011.05.011

Shidrang, S., Otte, M. and Bordes, J-G. (2020). The Yafteh cave, In: Smith, C. (ed.) *Encyclopedia of Global Archaeology*, Springer. DOI: 10.1007/978-3-319-51726-1_3270-1

Shidrang S. (2018). The Middle to Upper Paleolithic Transition in the Zagros: The Appearance and Evolution of the Baradostian. In: Nishiaki Y., Akazawa T. (eds), *The Middle and Upper Paleolithic Archeology of the Levant and Beyond. Replacement of Neanderthals by Modern Humans Series.* Springer, Singapore. DOI: 10.1007/978-981-10-6826-3_10

Shidrang, S., Biglari, F., Bordes, J-G. and Jaubert, J. (2016). Continuity and change in the Late Pleistocene lithic industries of the Central Zagros: a typo-technological analysis of lithic assemblages from Ghar-e Khar Cave, Bisotun, Iran, Archaeology. *Ethnology and Anthropology of Eurasia.* 44(1), 27-38. DOI: 10.17746/1563-0110.2016.44.1.027-038

Shidrang, S. (2014). Middle East Middle to Upper Paleolithic Transitional Industries, In: Claire Smith (ed), *Encyclopedia of Global Archaeology*, Springer, pp. 4894-4907. DOI: 10.1007/978-1-4419-0465-2_1855

Shidrang, S. (2009). A Typological and Technological Study of an Upper Paleolithic Collection from Sefid-Ab,Central Iran. XV Congress of the UISPP, book of abstracts, Vol.1, p.117-8, Lisbon. In: M. Otte, F. Biglari, and J. Jaubert (eds), *Iran Palaeolithic.* pp. 7-27, Proceedings of the XV World Congress UISPP, Lisbonne, Vol. 28, BAR International Series.

Zanolli, C., Biglari, F., Mashkour, M., Abdi, K., Monchot, H. Debue, K., Mazurier, A., Bayle, P., Le Luyer, M., Rougier, H., Trinkaus, E. and Macchiarelli, R. (2019). A Neanderthal from the Central Western Zagros, Iran. Structural reassessment of the Wezmeh 1 maxillary premolar. *Journal of Human Evolution* 135, 102643, DOI: 10.1016/j.jhevol.2019.102643.

## Professor Ravi Korisettar[3]

Professor Ravi Korisettar is a Senior Academic Fellow of the Indian Council of Historical Research, New Delhi. Ravi is a key contributor to Indian Palaeolithic archaeology, specialising in geoarchaeological methods and approaches to understanding the relationship between prehistoric humans and their environments in South Asia. He has published seven books in India and two abroad and is a Section Editor for *Current Science*, India's leading science fortnightly journal. Ravi has also held the position of Honorary Director of the Robert Bruce Foote Sanganakallu Archaeological Museum in Karnataka since its establishment in 2010.

### *What are your research interests and your particular area of expertise?*

In a couple of years from now, I will be completing fifty years of learning, teaching and researching in the field of archaeology. During the first half of this period, I experienced many ups and downs and institution-hopping to make a career as an archaeologist. This constrained me to work on diverse disciplines such as geoarchaeology, Quaternary geology, palaeoclimatology, radiometric dating, tephrochronology, the application of computer techniques, etc. This had enabled me to acquire multidisciplinary skills, though carrying the tag of 'jack of all and master of none' was a frustrating and sometimes depressing experience.

I am particularly interested in understanding man-land relationships in prehistory and explain why the settlements are found where they are. Currently, I am interested in global migrations and public outreach archaeology. Though primarily an archaeologist, I specialised in geoarchaeological field methods to address the problems of establishing the antiquity of Palaeolithic settlements, searching for hominin fossils, identifying refugia and critically assessing the correlation between climate and culture change.

### *What originally drew you towards archaeology?*

I was born into a low-income family. My parents used to inspire me with success stories about my Cambridge educated maternal uncle, S. Settar, about whom Raymond Allchin took pride in calling him a polymath. Though Settar was a historical archaeologist, his Cambridge experience had given him a clear interdisciplinary vision of archaeology. And yet, my parents wanted me to take up a combination programme of chemistry and physics for my undergraduate studies, which I completed in 1971. At this point, my uncle had returned from Cambridge and rejoined the faculty of Karnatak University. My brief visit to his place and a brief meeting with the Allchins at his residence was certainly a turning point for me.

Prior to this meeting, I used to spend my summer holidays in a small bookstore owned by my elder maternal uncle at Hosapete near Hampi, the well-known world heritage site in south India. The store used to sell fiction and scholarly works on art, tourism and culture. Tourists, students and scholars of Indian art and architecture from all over the world visiting Hampi used to drop by the bookstore and, though unable to speak fluently in English, I used to enter into conversation with some of them and also learn from them about other sites like Aihole, Badami and Pattadakal (also a World Heritage Site), the cradle of Indian temple architecture. This exposure to such books and scholars from all over, in addition

[3] Indian Council of Historical Research, New Delhi; korisettar@gmail.com

to the proximity of Hosapete to Hampi where we used spend our weekends, had given me some idea of what archaeologists do and I was also familiar with the adage that though the 'career of an archaeologist lies in ruins', it is full of romance and excitement.

During one of my conversations with Settar regarding the choice of a subject for post-graduation, it struck upon him that my science background will be helpful for an MA degree in archaeology. He advised me to go to Deccan College in Poona (Pune) and mentioned that great scholars like Iravati Karve, H.D. Sankalia and S.M. Katre, who have nurtured the disciplines of anthropology, archaeology and linguistics, built this world class institution. Undoubtedly, Poona had the reputation as 'Oxford of the east'. This was the most motivating advice I received at the critical point of my formative years of life and career. Coming to Poona changed my idea of archaeology and two years of post-graduate study brought me closer to appreciating Pleistocene geoarchaeology, bio-cultural evolution of man and to S.N. Rajaguru, a geologist by training. The latter's humble nature and exemplary attitudes drew me towards him and Pleistocene geoarchaeology and we built a lifelong relationship, academic and otherwise. Following my post-graduate studies, I enrolled for a PhD under his supervision.

■ "I became obsessed with identifying ashes," says archaeology professor Ravi Korisettar (in white) surveying Toba ash deposits at Jwalapuram in Andhra Pradesh. The Toba ash has been critical to understand human history and evolution.

## *What was your PhD topic? How did you find your PhD experience?*

The topic was Prehistory and Geomorphology of the Middle Krishna, a braided stream network draining the Precambrian basement complex on the Indian Peninsula. It was both exciting and frustrating. In the 1970s, discovering Acheulian artefacts was very rewarding and great material for writing a prehistory dissertation. However, I did not find any! The most exciting part of my research was my introduction to the works of Robert Bruce Foote, the father of Indian prehistory, and several other colonial and European geologists and geomorphologists. Stimulated by their works, I learnt the fundamentals of geology, fluvial geomorphology and climate change and became a competent field archaeologist. Fluvial deposits known as High Level Gravels were widespread in the Raichur and Shorapur Doabs in northern Karnataka (Doab refers to land between two rivers). The gravels were chiefly composed of chert (flint) clasts and the chert clasts were the chief raw material for making Middle Palaeolithic artefacts. These deposits and the associated Middle Palaeolithic chert artefacts were widely separated in time, provided little or no scope to determine the absolute timeline for human occupation of the area, other than 'Middle Palaeolithic age'. The absence of biotic material further hindered behavioural interpretation of hominins in a resource poor region Raichur Doab (land between the Krishna and Tungabhadra rivers). I became obsessed with the problem of the uneven distribution of Palaeolithic settlements across the subcontinent, their chronology and absence of hominin fossils.

### *After your PhD, what positions have you held and where?*

I did not have a permanent job for well over a decade after completion of my PhD in 1979. Short-term research assistantships kept me engaged in work, however did not promise a stable future and solid income. I was a Visiting Scientist (1980-82) at the Physical Research Laboratory in Ahmedabad where I was assigned the task of preparing a composite litholog of Neogene-Quaternary Karewa sediments in the valley of Kashmir and laboratory processing of samples for $Be^{10}$ dating, as well as process samples for palaeomagnetic, micropalaeontoligic and palynologic analyses. Following this I had a post-doc fellowship (1983-85) from the Indian Council of Historical Research (ICHR), New Delhi, and a research associateship (1988-89) at Deccan College, with intervening unemployed days. Finally, I was appointed Reader (1989-98), then eventually Professor (1998-2013), at Karnatak Univeristy. Post retirement, I was Dr. DC Pavate Chair Professor (2013-15) at Karnatak University, Dr VS Wakanakr Senior Fellow (Bhopal, Madhya Pradesh), UGC Emeritus Fellow (2015-17) and now I am concurrently ICHR Senior Academic Fellow (2019-21), Adjunct Professor at National Institute of Advanced Studies, Bengaluru (since March 2020) and Hon. Director of the Robert Bruce Foote Sanganakallu Archaeological Museum at Ballari in Karnataka (since 2010).

### *What current projects are you working on? Where do you hope these go in the future?*

I have multiple projects on going: (a) understanding the cognitive content of prehistoric rock art, (b) re-examining of Late Pleistocene hominin fossils from rock shelter excavations and assessing their potential for aDNA studies in collaboration with Sheela Athreya, Texas A&M University, USA, (c) preparing systematic catalogue of antiquities from surface surveys and excavations carried out during the last forty-five years, now handed over to the government at the Robert Bruce Foote Sanganakallu Museum Archaeological museum in Ballari, Karnataka, and (d) preparing a comprehensive report on Sangankallu Neolithic-Iron Age excavations.

My publications bear ample testimony of my many successful international collaborations and that I will be able to successfully carry out these research projects and contribute to a better understanding evolution of past human societies in a multidisciplinary framework in the future.

### *What project or publication or discovery are you most proud of?*

I am very proud of the following achievements:

1. The discovery of tephra marker bed in the alluvial sediments of the Indian Peninsula (Korisettar et al. 1989).
2. The development of a Basin model to delineate man-land relationships (Korisettar, 2007).
3. The establishment of Robert Bruce Foote Sanganakallu Archaeological Museum (2020).
4. The first dating of the Middle Palaeolithic and the oldest date for the microliths in India (at the time of publication, 2009)
5. The emergence of agricultural economies in the Southern Neolithic of India (chief investigator Dorian Fuller now at UCL, London)
6. The Bellary District Archaeological Project (Co-investigator: N.L. Boivin, now at Max Planck Institute, Jena)
7. The Kurnool District Archaeological Project (Co-investigator: M.D. Petraglia, now at Max Planck Institute, Jena).

*Replicas of human ancestors at the Robert Bruce Foote Sanganakallu Archaeological Museum*

### *What are your favourite memories of your career?*

Memories have been both sweet and sour, but more on the sweeter side. The early decades of my archaeological career was a period of anxiety and stress, compounded by not being able to contribute to the growth of archaeological knowledge through my PhD work. Job applications to the Archaeological Survey of India and several institutions did not find me suitable because of 'other considerations'...

My entry into Karnatak University in 1989, though helped me breathe a sigh of relief, moved me away from full time research in archaeology to full time teaching in history and archaeology, where archaeology was a subsidiary component of postgraduate syllabus. During my harness at the university, I continued to confront non-egalitarian environments, both socially and academically. Yet winter and summer holidays were at my disposal to pursue my research interests and update myself with the developments in method and theory in global prehistory. During the settling in time of a year or so, I began to explore the scope of interdisciplinary collaborative research with scholarly friends from institutions in India and abroad.

The Ancient India and Iran Trust- Charles Wallace fellowship (1996) at Cambridge, UK, gave me my first international exposure to intense and stimulating academic experience. The Fulbright Visiting Scholarship (2001) at the Smithsonian Institution in Washington DC gave me greater international visibility and strengthened my wide network with archaeologists in India and abroad.

*A down-scaled model of Sanganakallu Neolithic hills at the Robert Bruce Foote Museum.*

### *If you were not an archaeologist, what would you be?*

Archaeology was my bread winner. If I were not an archaeologist, I would have to opt for an undergraduate lectureship (if considered suitable, of course) or turn towards local industry for a non-academic job.

### *What advice would you give to a prospective student interested in your field of research?*

Though there have been great leaps in Indian archaeology, especially in the areas of Palaeolithic and Neolithic, I see that Indian archaeology is more productive since the turn of the century. The application of processual and post-processual archaeological methods and theory have opened up new pathways of investigation aimed at holistic reconstruction of human bio-cultural and social evolution. Our priorities are the issues relating to (a) identifying potential sites for geochronology of Palaeolithic sites, (b) reconstructing palaeogeography of Palaeolithic landscapes for a better understanding of site formation processes, (c) delineating man-land relationships during the Quaternary, (d) developing ethnoarchaeological interpretations of archaeological data sets and (d) helping place the Indian subcontinent at the forefront of global debates on peopling of the earth. So, I would advise students to concentrate on these topics.

*Ravi engaging in public outreach archaeology, with school children at Jwalapuram.*

### *If you had a time machine, how far would you ask to go back, where would you go, and what would you want to see?*

I would be at the time of Big Bang, witness the formation of the atmosphere and the origins of first life forms and travel with the emergence of multiple life forms. Then I would also witness the emergence of hominins capable of making and using tools. It is a fantasy though, of course!

### *References:*

Korisettar, R. (2007). Towards developing a basin model for Palaeolithic settlement of the Indian subcontinent: Geodynamics, monsoon dynamics, habitat diversity and dispersal routes. In M.D. Petraglia and B, Allchin (eds), *The Evolution and History of Human Populations in South Asia*, Springer, pp. 69-96.

Korisettar, R., Venkatesan, T. R., Mishra, S., Rajaguru, S. N., Somayajulu, B. L. K., Tandon, S., Gogte, V. D., Ganjoo, R. K. and Kale, V. (1989). Discovery of a tephra bed in the quaternary alluvial sediment of Pune district (Maharashtra) Penisular India. *Current Science* 58(10), 564-567. ISSN 0011-3891

## Dr Jennifer French[4]

Dr Jennifer French is a Lecturer in Palaeolithic Archaeology in the Department of Archaeology, Classics and Egyptology at the University of Liverpool. Jennifer completed her PhD at the University of Cambridge, followed by a Research Fellowship in Archaeology and Anthropology. She then undertook a postdoctoral position at University College London as a Leverhulme Trust Early Career Fellow and then as a Wenner-Gren Hunt Postdoctoral Fellow. She has just finished writing a monograph titled 'Palaeolithic Europe: A Demographic and Social Prehistory' (to be published by Cambridge University Press in their World Archaeology Series). This will be the first comprehensive synthesis of the population history of the European Palaeolithic combining archaeological data with osteological, genetic, and ethnographic data.

### *What are your research interests and particular area of expertise?*

I'm a Palaeolithic archaeologist with particular expertise in the European Middle and Upper Palaeolithic (Neanderthals and the first *Homo sapiens* in Europe). Within Palaeolithic archaeology, I'm very much a generalist, and my research is centred around key themes, rather than focused on the analysis of any one particular class of material. These themes are: archaeological demography, the theoretical and methodological challenges of the archaeology of archaic hominins, and the integration of Palaeolithic archaeology with the wider anthropological sub-field of hunter-gatherer studies. I like to think that my research programme bridges divisions between the scientific nature of human evolutionary studies and the humanistic focus of prehistory.

### *What first inspired your interest in archaeology? Did you always want to be an archaeologist and/or academic?*

I am not one of those archaeologists whose career is the realisation of a long-held childhood ambition. I was, however, lucky enough to take my A-Levels at a large FE college- one of the few in the UK that offered the (sadly, now discontinued) Archaeology A-Level. I was uncertain what to choose for my fourth A Level subject (to study alongside German, English Literature, and Theatre Studies), and selected Archaeology because of the breadth of the subject. Within one week of classes, I was absolutely hooked, and haven't looked back since.

I was even later in realising that I wanted to be an academic, mostly because I didn't know that 'academic' was a career until towards the end of my undergraduate studies. I was the first in my family to attend university, and honestly hadn't thought much about what I would do once I graduated beyond probably continuing in my job at a supermarket while I assessed my options. I had surprised myself by how well I had done in my undergraduate studies and how well received my work had been, so I decided that my preferred next step was graduate study. I was lucky enough to get a full scholarship from St John's College, Cambridge to study for my Masters' degree in Archaeological Research. It was then that I decided to pursue a career as an academic as long as academia would have me!

---

[4] Department of Archaeology, Classics and Egyptology, University of Liverpool, U.K.; jennifer.french@liverpool.ac.uk

*Excavating in Bulgaria in 2006 as an undergraduate at Durham University.*

***Where did you complete you PhD, what was your topic and who was your supervisor?***

I stayed at Cambridge for my PhD, completing a thesis entitled 'Populating the Palaeolithic: a Palaeodemographic Analysis of the Upper Palaeolithic Hunter-Gatherer Populations of Southwestern France' funded by the AHRC under the supervision of Prof. Paul Mellars, who continues to be a close friend and mentor. This thesis was a continuation of my Masters' research into population changes across the Middle-Upper Palaeolithic transition in the same region, the results of which I published with Prof. Mellars in Science in 2011 (Mellars and French, 2011). Both of these projects looked at how a range of archaeological data can be used as proxies for relative changes in past population size and density of Palaeolithic hunter-gatherer communities, and the relationship(s) between these documented demographic changes and changes in other (environmental, social) domains.

*Visiting the famous 'Lion Man/ Löwenmensch' in Southwest Germany while a PhD student at the University in Cambridge.*

***What were the main findings from your PhD? Have you done any further work on this since you completed your PhD?***

From the perspective of understanding the Upper Palaeolithic of Southwestern France, the main result of my PhD was the construction of a temporal sequence of fluctuations in relative population size across the Upper Palaeolithic spanning the Aurignacian-Azilian technocomplexes. Some of these fluctuations I didn't see the relevance of at the time- for example, it took my colleague Andreas Maier finding a similar demographic trough in the Late Gravettian of Europe more widely (Maier and Zimmerman, 2017) for me to realise that this hadn't really been documented before and goes against prevailing narratives that it was during the LGM 'proper' that the Palaeolithic population of Europe was at its smallest. From the perspective of archaeological demography, my PhD research also tested the robustness of palaeodemographic proxies. Along with Christina Collins, I demonstrated that the proxies of archaeological site counts and summed probability distributions of 14C dates produce similar demographic patterns for the French Upper Palaeolithic record- a finding that has since been replicated elsewhere and increases our confidence in the demographic signature produced by these proxies.

I recommend that those particularly interested in the results of my PhD check out the three papers that resulted from this research in *Journal of Anthropological Archaeology* (French, 2015), *Journal of Archaeological Science* (French and Collins, 2015) and *Journal of Archaeological Theory and Method* (French, 2016).

As I discuss further below, demography has continued to be the focus of my post-PhD research. In addition to my main projects, I co-lead a working group on 'Cross-Disciplinary Approaches to Prehistoric Demography' (CROSSDEM) with colleagues at the universities of Bournemouth, Barcelona, and Alicante. Collectively, we have hosted several international workshops, and we are putting the finishing touches to a CROSSDEM Special Issue of 'Philosophical Transactions of the Royal Society B', which brings together a series of 'state of the art' papers at the forefront of research in prehistoric demography. I've also developed an approach that integrates the study of demography with the study of women and of gender in prehistoric contexts (French, 2018).

***Where have you worked since completing your PhD? On what projects?***

I was very lucky in that my first post-doctoral position overlapped with my PhD. I took up a Research Fellowship in Archaeology and Anthropology at Peterhouse, Cambridge in 2012, and spent the first 6 months of that post completing my PhD thesis. I remained at Peterhouse until 2016 publishing the results of my doctoral research as discussed above and planning for my next major project which I carried out at the UCL Institute of Archaeology, first as a Leverhulme Trust Early Career Fellow and then as a Wenner-Gren Hunt Postdoctoral Fellow.

This project 'Palaeolithic Europe: A Demographic and Social Prehistory' weaved together archaeological, palaeoanthropological, and genetic data, alongside ethnographic data on recent foragers and demographic models of extant small-scale societies, to develop a demographic prehistory of European Palaeolithic populations ~1.8 million to 15,000 years ago. The results of this project are presented in my monograph of the same name, which I have just submitted to my editor at Cambridge University Press. This book advances a novel structure for examining the European Palaeolithic based around four demographic stages; 1) visitation; 2) residency; 3) expansion, and; 4) intensification. By combining evolutionary frameworks (Human Behavioural Ecology, Life History Theory) with a social and gender-aware approach to investigating Palaeolithic societies, this book addresses both the biological and social drivers of demographic change within and between hominin species and populations, refuting long-standing ideas about the stability of the demographic regimes of Pleistocene hunter-gatherers.

***You are just starting your new position as a Lecturer in Palaeolithic Archaeology at the University of Liverpool. Please tell us a little bit about what you are hoping to bring to our department and what you have planned for this role.***

Firstly, I would like to say how thrilled I am to be joining the Department of Archaeology, Classics, and Egyptology, and especially the Archaeology of Human Origins research group at the University of Liverpool. I'm particularly excited to join a department with such a strong emphasis on teaching Palaeolithic Archaeology and Evolutionary Anthropology at the undergraduate level, and one of the things I hope to bring to the department is a continued commitment to sharing the latest research and ideas with students. I also have lots of potential undergraduate and masters' dissertation projects that derive from my recent book research, so if you are a student at Liverpool interested in Middle-Upper Palaeolithic archaeology, hunter-gatherer studies, or archaeological demography, please get in touch!

My research plans for the next few years are varied and given the lottery-nature of many funding schemes, I don't want to jinx them! Nonetheless, I hope to collaborate with some of my more quantitatively minded new colleagues on some demographic questions that have been vexing me for years, as well as continuing my work with Prof. April Nowell at the University of Victoria on adolescence in the Palaeolithic (Nowell and French, 2020). I'll also be continuing my fieldwork at several Palaeolithic cave sites in the UK with my colleague, Dr Rob Dinnis of the University of Aberdeen.

*With the excavation team at Kents Cavern, Devon. Photo credit: Rob Dinnis.*

***What do you think has been the most revolutionary discovery in Palaeolithic archaeology over the past 5 years?***

I'm a big advocate for the position that 'archaeology is not about what you find, it's about what you find out' but there's no denying that there have been some pretty spectacular discoveries in Palaeolithic archaeology over the last 5 years! As someone whose research focuses on the dynamics of European Palaeolithic populations, the possibility that *Homo sapiens* were present on the continent as early as ~210,000 years ago really throws a spanner into our models and assumptions about the correlation between hominins and different lithic industries in the European Palaeolithic (Harvati *et al.*, 2019). The finding of a child with a Neanderthal mother and a Denisovan father is also incredibly exciting from a demographic perspective-the notion that we now have direct evidence for this sort of interaction during early prehistory is mind-blowing (Slon *et al.*, 2018).

### *What are your favourite things about being an academic? What would you change?*

There are many things that I love about being an academic: the freedom to pursue interesting research questions, to work with great colleagues and teach students, and to feel that you are contributing to a global knowledge base. The bad things and the things I would change are well-documented: the increasing casualisation of the academic work force and associated structural problem in terms of the opportunities afforded to people of different genders, nationalities and ethnicities. These issues are obviously not unique to academia and academic settings, but I do think we have a greater responsibility within the academy to be a force for positive change in these areas. I have a long track-record of public engagement endeavours and initiatives to get more young people into both higher education and archaeology (see for example, my work on the National University Archaeology Day with colleagues from the UCL Institute of Archaeology), but there is much, much, more to do here.

*Teaching field techniques at Ffynnon Beuno, Wales. Photo credit: Rob Dinnis.*

***References:***

French, J. C. (2015). The demography of the Upper Palaeolithic hunter–gatherers of Southwestern France: A multi-proxy approach using archaeological data. *Journal of Anthropological Archaeology* 39, 193-209. DOI: 10.1016/j.jaa.2015.04.005.

French, J. C. (2016). Demography and the Palaeolithic Archaeological Record. *Journal of Archaeological Method and Theory* 23, 150-199. DOI: 10.1007/s10816-014-9237-4.

French, J. C. (2018). Archaeological Demography as a Tool for the Study of Women and Gender in the Past. *Cambridge Archaeological Journal* 29(1), 141-157. DOI:10.1017/S0959774318000380.

French, J. C. and Collins, C. (2015). Upper Palaeolithic population histories of Southwestern France: a comparison of the demographic signatures of 14C date distributions and archaeological site counts. *Journal of Archaeological Science* 55, 122-134. DOI: 10.1016/j.jas.201.01.001

Harvati, K., Röding, C., Bosman, A.M., Karakostis, F.A., Grün, R. Stringer, C., Karkanas, P., Thompson, N.C., Koutoulidis, V., Moulopoulos, L.A., Gorgoulis, V.G. and Kouloukoussa, M. (2019). Apidima Cave fossils provide earliest evidence of *Homo sapiens* in Eurasia. *Nature* 571, 500–504. DOI:10.1038/s41586-019-1376-z

Maier, A. and Zimmermann, A. (2017). Populations headed south? The Gravettian from a palaeodemographic point of view. *Antiquity* 91(357), 573-588. DOI:10.15184/aqy.2017.37

Mellars, P. and French, J. C. (2011). Tenfold population increase in Western Europe at the Neandertal–to–Modern Human Transition. *Science* 333(6403), 623-637. DOI: 10.1126/science.1206930

Nowell, A. and French, J. C. (2020). Adolescence and innovation in the European Upper Palaeolithic. *Evolutionary Human Sciences* 2, E36. DOI:10.1017/ehs.2020.37

Slon, V., Mafessoni, F., Vernot, B., de Filippo, C., Grote, S., Viola, B., Hajdinjak, M., Peyrégne, S., Nagel, S., Brown, S., Douka, K., Higham, T., Kozlikin, M.B., Shunkov, M.V., Derevianko, A.P., Kelso, J., Meyer, M., Prüfer, K. and Pääbo, S. (2018). The genome of the offspring of a Neanderthal mother and a Denisovan father. *Nature* 561, 113–116. DOI: 10.1038/s41586-018-0455-x

## Dr Rebecca Wragg Sykes[5]

Dr Rebecca Wragg Sykes is a Palaeolithic archaeologist, Honorary Research Fellow of the University of Liverpool and author of new popular science book: Kindred. Following her PhD at the University of Sheffield where she worked on analysing the evidence for late Neanderthals in Britain, in 2013 she won a prestigious Marie Curie postdoctoral fellowship at the PACEA laboratory, Université de Bordeaux. Following that, she has been working largely outside of scientific research, nurturing projects in creative heritage consultancy and popular science writing as well as co-coordinating TrowelBlazers as part of her advocacy work to improve equality in archaeology. Her most recent project has been the writing of her first book titled Kindred: Neanderthal Life Love Death and Art, published with Bloomsbury Sigma, which explores our ever-evolving understanding of Neanderthals and their culture. Already a bestseller, Kindred has been critically acclaimed in *Science* and *Nature,* widely reviewed in the media (listed as one of the 'Best Books of 2020' by the Times) and has been translated into sixteen language so far.

***What are your research interests and particular area of expertise?***

I'm interested in all of the Palaeolithic, however I've homed in specifically on the Middle Palaeolithic and Neanderthals. Funnily enough, what had originally attracted me to the University of Bristol for my undergraduate degree was that they had an Upper Palaeolithic rock art course, but my research ended up going back further in time. Whilst the Upper Palaeolithic is abundant and interesting, I like the challenge of the Middle Palaeolithic as you have less evidence to deal with. Plus, assemblages with tiny bladelets scare me...!

I'm primarily a stone tool person and was trained in lithic analysis. During my masters at Southampton, there was an awesome teaching collection based on 19th century sites which fostered my interest in Middle Palaeolithic stone tools. I decided to study the Middle Palaeolithic of Kents Cavern for my Master's project. But, although I love lithics, I'm interested in all aspects of Neanderthal life. The quality and breadth of the data that we now have allows us to explore the interconnections between these different aspects to really understand our not-so-distant relatives. I really like looking at interconnections - something I learned from my supervisor at Southampton, Professor Clive Gamble!

***What originally sparked your interest in human evolution studies?***

I have always been interested in the past. I am one of those cliché archaeologists who dug up pot-shards in their back garden and collected dead creatures as a child! As a family, we went on many holidays to historic sites so my desire to imagine the past was always there. Ultimately, I had to choose between going to art college or to do archaeology at university and, though neither of them had great career prospects, I thought I might have a slightly better chance at making a living as an archaeologist!
In terms of prehistory, like many people, I was somewhat attracted to the mystery and that deep-time connection to our ancient human past. I'm not ashamed to admit that I was a fan of *The Clan of the Cave*

[5] Department of Archaeology, Classics and Egyptology, University of Liverpool; rwraggsykes@gmail.com

*Bear* books when I was a teenager; people can critique the story but I was totally absorbed by Jean Auel's descriptions of the environment and how prehistoric people lived. I didn't understand what she meant by 'striking platform' at first, but I eventually worked that one out! As I have found in writing Kindred, describing lithic technology to a lay-audience is really hard without a lot of visuals. Considering that, her books are very impressive and she took a lot of time to research them, so they're definitely something that has stayed with me.

### *What was your PhD topic? How did you find your PhD experience?*

My PhD was the first full analysis of the British Mousterian, the lithic culture created by late Neanderthals. As part of my project, I also addressed the faunal record and landscape in relation to these lithic assemblages. Prior to my PhD, little work had been conducted over the last 20 years on this material, despite there being many shifts in the chronological frameworks adopted and how we consider and analyse stone tool assemblages. I used techno-economic methodologies in my analysis, moving away from typological assessments as has typically been done previously, as I wanted to bring a coherence and consider the assemblages as a whole, despite there not being a lot of material. Compared to continental sites, the British record is minuscule, but this doesn't make it useless. That said, we do need to be extremely careful as many British assemblages were excavated very early using entirely different excavation and recording practises; the only British late Middle Palaeolithic site that has been excavated to modern standards is Lynford Quarry. I studied the main assemblage at this site which was fascinating.

My PhD was very data collection intensive as a result of early British excavators who had a philanthropic tendency of wanting to distribute their collections across different museums. Because of this, I had to go to a huge number of museums across the UK, sometimes to only study three artefacts! The assemblages were so tiny so I wanted my analysis to be as comprehensive as possible, only missing a few collections in the US and Cork. I really enjoyed this aspect of my PhD, especially the independent study and getting an idea about the realities of archaeological collections. I also had a great time with my fellow PhD cohort at the University of Sheffield. I was one of two studying the Palaeolithic, but through my other friends was exposed to ideas about later prehistoric and historical archaeology. I found this really benefited my development as an archaeologist as, before I started my PhD, I was very much a part of the scientific-positivist trend in archaeology. I found being exposed to a wider theoretical approach at Sheffield challenging but I enjoyed the broadening out a lot. During my PhD, I set up a journal discussion group at Sheffield with my fellow research students and I edited the student peer-reviewed journal *Assemblage*, which was great.

### *After your PhD, what positions have you held and where?*

After the PhD, I didn't have much luck getting positions, like a lot of people. In total since my PhD, I applied to 20 positions and I got 1 interview offer, which was difficult. I was eventually awarded the Marie Curie post-doctoral fellowship in 2012, which I didn't start until June 2013. I ranked sixth in the reserve list of the thousands of applicants and so I didn't find out I had been awarded the fellowship until months afterwards, by which time I had already prepared myself for not getting it. It was a bit of a shock! I moved to France in 2013 and stayed there for four and a half years. The fellowship was only two years (which was super intensive) but I had a child, so my maternity leave extended the postdoc a little, plus I stayed over there for a while afterwards applying for more positions and writing Kindred. My postdoc was great because although it was really challenging moving abroad and into a completely different research culture, I had some French language ability which made it a little easier.

*Silcrete from Saint-Pierre-Eynac quarry excavations.*

The project was focussed on looking at landscape-scale lithic sourcing and connectivity between sources and sites. The site I was excavating was a silcrete quarry in Massif central, a material source site used throughout prehistory by Neanderthals and later populations. We surveyed the hill where the source site was believed to be and, right at the top, there was an immense spread of knapping debris. So, we hit the jackpot right away! However, this particular silcrete is not easy to analyse as it has strange fracture planes, which makes it difficult to differentiate which rocks were modified by humans vs. natural processes. It took me a while to get my head around it. Also, it was clearly a mixed age quarry site, which made sampling very difficult. We took some surface samples and dug some test-pits, where lower down we found evidence for Middle Palaeolithic Neanderthal presence, which was great. What was also really nice was that, although it was primarily a silcrete site, when we did the surface survey we found two tiny flint artefacts which were directly sourced from far south in Ardeche, France (Wragg Sykes et al., 2017).

*TRACETERRE Marie Curie fieldwork team: Dr Jean Paul Raynal, Vincent Delvigne, Rebecca Wragg Sykes, Erwan Vayssié.*

After my postdoc I again applied for a lot of positions but, eventually I decided to stop since the thought of potentially moving to a third country was too much with a young family and Brexit on the horizon. Our only option was to come back to the UK, where I've been focussing on writing my book, alongside other creative and writing projects, plus research consultancy work, including for the National Trust, for the two and half years since.

***Tell us a little bit about your book 'Kindred'. What is the book about, why did you want to write a book about this topic and how did you find the book writing process?***

Kindred is an up-to-date definitive book about the Neanderthals which is accessible to all. Whilst it is not an academic text, it draws on my academic expertise and dives deeply into selected archaeological sites. I wanted to show people how mind-blowing 21st century archaeology is, exactly what we can do and how we connect across different subdisciplines. I've tried to show the aspects of the archaeology that are fascinating but never seem to get into the headlines which have informed the revolution of how we understand Neanderthals over the last 30 years. Lithics aren't easily explained by media stories! I've tried to make it accessible for everyone and something useful even for academics, particularly for students who have no background in Neanderthals or those who study prehistory but want an overview of the current debates, evidence, knowledge and the types of data. The book does have some narrative exploration, which of course I would never be able to put into academic papers. It has been wonderful to let the creative side of my writing flourish and develop. I really hope my colleagues find the book interesting!

*Kindred jacket.*

I found it very different to writing an academic paper, though the amount of literature reviewed was vast. I chose not to include a bibliography in the book as I didn't want to put lay-people off, but I am doing an online version on my website which is so far over 120 pages long. It's huge because everything I wrote has been informed by the literature. I went through the detail historically as well as looking deeply at individual sites, following the evidence sometimes right through to unpublished reports. This was very time-consuming, but I wanted to make sure I did it properly. It was also nice not to have to do in-text citations! I found the process was difficult in another way in that I had to restructure the book twice. I didn't want it to be chronological as that has been done many times before, so for me it was always going to be thematic. However, the interconnectedness between themes made it very difficult to structure as, for example, it is hard to talk about diet without talking about social cooperation etc. So that was challenging. In terms of the creative aspects of the book, I always wanted to do that right from the beginning, but I wasn't 100% sure at first. After a while, it started really flowing nicely and I think it works well. There may even be a couple of poems in there...

Overall, it's been a really fun experience and the response from everyone has been absolutely overwhelming! It's incredibly gratifying that everyone seems to be responding so well to exactly the things the book intended to do, showing people the details as well as the difficulties archaeologists face when dealing with this material to make informed inferences about the deep past. I had hoped to connect and engage people emotionally with Neanderthals and their culture, and it seems to have worked, I think!

***What projects are you currently working on to advance equality in archaeology?***

Equality within archaeology has been interesting to me for a long while, but my involvement in these movements largely came about during my postdoc. I was connecting with some other early career researchers on Twitter (Brenna Hassett, Suzanne Pilaar Birch and Tori Herridge) and on there we discussed how there was a lack of open-access public-facing resources about women in archaeology, leading to the idea for TrowelBlazers. It evolved from being a Tumblr account to a professional website. Since we built the website in 2015, it has changed a lot as the vast majority of our material is now sourced from the community. People have really responded to what TrowelBlazers represents which is amazing to see. For me personally, I love the creative side of this project. Just after finishing my postdoc in October 2015, TrowelBlazers was approached by the artist Leonora Saunders with a really cool idea to produce a photographic exhibition. I project managed this exhibition (Raising Horizons) for TrowelBlazers as I was the only one not fully employed at the time. I selected and researched all of the women included in the exhibition and, together with Leonora, came up with the objects they would hold and the settings. The whole process took well over a year and required a massive CrowdFund to fund it, which was amazingly popular. Being involved with the exhibition took me back to my early days at school when I was doing art; it was so lovely to be able to be involved with something like that, which would have been impossible in a scientifically structured full time career.

Since then, I have continued to work on the advocacy side of things. We have joined forces with multiple other groups in the UK interested in improving diversity and equality, recently forming the IDEAH federation (Inclusion, Diversity and Equality in Archaeology and Heritage). We were going to start meeting and working on projects before COVID-19, though hopefully we can reanimate that and restart as a group amplifying each other's voices soon. Our aim is to look at how we can practically address issues not only in the professional sphere but also in universities, and we have a lot of ideas about how we can encourage university departments to take proactive steps in terms of equality, representation and harassment. Archaeology has a long way to go in this regard, so I'm really pleased to carry on this work.

***What is the best thing about your work? What is one thing that you would change if you could?***

The best thing about my current situation is the freedom. I do miss doing academic research and working with materials. I haven't been digging for a long time. However, I haven't disconnected from intellectual research and writing the book has really helped with that. I also have some papers simmering and academic projects that are ongoing, and I was recently invited to contribute a chapter to an upcoming Oxford handbook on cognitive archaeology. So, I feel like I have maintained my connection to scientific research despite not having an academic position. Twitter has really helped me stay in active discussion with the research community too, which has been a godsend as working at home can be quite isolating. I do have a lot of freedom to pursue what I want which I love. But in terms of what my next step will be, I'm not quite sure though I have some ideas...

If I could change something, it would be the lack of reliable income! I also wish that academia would be more open to proper part-time roles and flexible working, as not everyone who wants a research career can (or wants) to commit to a full-time post.

*Raising Horizons exhibition launch at The Geological Society, London, with Lenora Saunders.*

### *What advice would you give to a student interested in getting into your field of research?*

For a student interested in working on Neanderthals from an academic perspective, I would say learn German! French is great but there are a lot of active projects in Germany right now. Germany seems to be one of the best places for funding and, whilst I don't want to put a downer on things, if you are a student in Britain right now, things aren't looking great in that regard. I really hope new research structures are going to evolve to make it possible to maintain international collaborations. That is one of the most precious things about all of the experiences I've had, working with different people from different backgrounds and cultures. This is key to human origins. Interconnection between different countries is vital, though we don't see enough between Europe and the Global South currently. I would say to have an open mind about that!

In a broader sense and assuming not every student will get an academic position, as I hadn't for a long time, I would ask them what is it about archaeology that makes them like it, and can they take these skills and apply them to different roles. Don't be afraid of shifting disciplines or professions and be flexible! It's okay to have a career more like a braided river than a path.

***References:***

Wragg Sykes, R. M., Delvigne, V., Fernandesa, P., Piboule, M., Lafarge, A., Defive, E., Santagata, C., Raynala, J.P. (2017). 'Undatable, unattractive, redundant'? The Rapavi silcrete source, Saint-Pierre-Eynac (Haute-Loire, France): Challenges studying a prehistoric quarry-workshop in the Massif Central mountains. *Journal of Archaeological Science: Reports* 15(2017), 587-610. DOI: 10.1016/j.jasrep.2017.07.022

## Professor Nena Galanidou[6]

Professor Nena Galanidou is a Palaeolithic archaeologist at the University of Crete. She obtained her PhD in Palaeolithic Archaeology from the University of Cambridge in 1996, where she later became a research fellow until 1999. Since 2000, she has been teaching Prehistoric Archaeology at the University of Crete. She has conducted fieldwork in Greece, Croatia and Israel. Nena has also participated in international projects studying the Palaeolithic and Mesolithic of southeast Europe and directs Palaeolithic research on the island of Lesbos, excavating the Lower Palaeolithic Lisvori-Rodafnidia open-air site, and in the Inner Ionian Archipelago, excavating the Middle Palaeolithic Panthera Cave on the islet of Kythros.

### *What are your research interests and your particular area of expertise?*

I am a Palaeolithic archaeologist currently working on three thematic areas: the Acheulean, the Middle Palaeolithic puzzle, and Continental Shelf Prehistoric Research. My early work on Spatial Archaeology and Hunter-Gatherer Ethnoarchaeology reflects two more research interests that are always alive and sparkling.

### *What originally drew you towards human evolution studies?*

It was this internal need to explore the human condition. Upon making a career decision, I chose to leave aside the wonders of Greek archaeology, a siren that I closed my ears to, and opt for the beauty of Palaeolithic archaeology. Also, for a while I oscillated between my penchant for maths and my passion for the past. I hold an MSc in Archaeological Computing that gave me a permanent job at the Benaki Museum at the heart of my beloved city, Athens, but I gave it up to pursue around the world my true love, the archaeology of human evolution.

My point of departure was a humanistic view of world history. Through my research, I wanted to take a leap, go beyond the differences and reach our deep roots to those things that unite humans as a species, as a common heritage before today's national, linguistic, religious, class, race or gender differences. Of all archaeological specialisations, the archaeology of human evolution offers its practitioners worldviews that are planetary (think globally act locally) and tolerant (we humans are diverse yet fundamentally united through a shared past and common threads such as genes, evolutionary habits and technological innovations).

In due course I came to realise the excitement that comes from Palaeolithic work in the field or the lab. Human evolution is a book that is continuously being re-written. New finds and new readings of old finds shed new, often unexpected, direct or oblique light on the old threefold question: who are we– where do we come from – how did we get here. Whether it is a submerged cave, like Cosquer, a lion pack depicted in the Chauvet Cave, a small *Homo sapiens* fossil bone from the Misliya Cave lady in the Levant, a jaw bone belonging to a Denisovan from the Baishiya Cave on the frigid Tibetan Plateau, the hidden hearths of GBY that speak eloquently of fire mastery at the onset of the Middle Pleistocene, the hibernating Atapuercans, or the Lomekwi tools that push Palaeolithic beginnings further back in deep

[6] Department of History and Archaeology, University of Crete; galanidou@uoc.gr

time, all are landmarks showing that we will never be bored of palaeoanthropological discovery and debate. The canon of our field is a fast-changing one, a bit like iPhone models, there is always a new launch coming soon.

*Tabun Cave excavation, lunch break above the site, February 2020.*

**What was your PhD topic? Where did you complete your PhD and who was your supervisor? How did you find your PhD experience?**

My topic was the Upper Palaeolithic use of space in cave sites, and my supervisor was Geoff Bailey in the Department of Archaeology, Cambridge University. In this work I brought together three themes I was fond of: hunter-gatherer archaeology, architecture and mathematics. Using archaeological material from Epirus and Bosnia in southeast Europe, I examined the origins of architecture: in other words, if and when some structure is identified in an otherwise unstructured space. The caves offered a natural shell for protection, which varied as to the area available or the constraints present. The study employed an array of statistical methods to map the distribution of finds and concluded that in this early use of space, hearths acted as the primary cohesive elements in the spatial organization and activities of the social group (Galanidou, 1997).

The social milieu at Cambridge during the 1990s was very international, certainly thought-provoking, sometimes lonely and more often exciting. Little has changed since then; it is the sort of place that begets innovation through a very stable annual routine. My tempo was as follows: long study winters in the UK (sometimes even up to June, when college heating was turned off no matter what the thermometer suggested), interspersed with short summers conducting fieldwork in Greece. This formation period endowed me with some of my lifetime friends, a taste for Scandinavian architecture and furniture, as my college, Clare Hall, was designed by Pritzker prize winner Ralph Erskine, and a preference for single malt. At Cambridge, I loved the Haddon Library and its friendly staff but could not stand the freezing temperatures of the University Library stalls. I cycled everywhere no matter the weather and loved the flea markets at the town hall and the live gigs at the Corn Exchange. Beyond my archaeological formation, I am grateful to the Cambridge ecosystem, for it gave me some superior lessons in British diplomacy.

### *After your PhD, what positions have you held and where?*

Between 1996 and 1999, I held a postdoctoral fellowship at Clare Hall, Cambridge that gave me the opportunity to do research, publish and teach without the worry of having to earn a living. During that time, I taught Quantitative Methods in Archaeology and Mesolithic Archaeology courses to the demanding audiences of Cambridge undergraduates in the Archaeology Department.

In 2000 I got a lectureship and since then I have climbed up the professorial ladder in the Department of History and Archaeology of the University of Crete, Greece.

*...Still digging with a broken wrist (!) Panthera Cave, October 2020.*

### *What current projects are you working on? Where do you hope these go in the future?*

My current projects study the archaeology of the Lower and Middle Palaeolithic in Greece through a common denominator, Island Archaeology, targeting islands of different scale, geography and resources.

On the island of Lesbos, in the northeast Aegean Sea, my team is exploring the first extensive Acheulean settlement in southeast Europe and western Anatolia (Galanidou *et al.*, 2016). On the banks of small rivers and the shore of what was a big palaeolake, the Kalloni Gulf, and against a volcanic setting, we are unearthing a cluster of stratified Lower Palaeolithic sites, placing Greece on the map of the Acheulean world. Our finds link the Middle Pleistocene archaeology of the Aegean with the corresponding archaeology of Africa and Eurasia, and underline the importance of volcanic geographies. The principal site of Lisvori-Rodafnidia is situated by a thermal spring, as are other Acheulean findspots on Lesbos. From our work a new scenario emerges for the early colonization of Europe: at least half a million years ago, hominins walked into Europe via the Aegean Region, during periods of low sea level stands, following tracks that certainly Early Pleistocene animals and perhaps hominins had followed too (Sakellariou and Galanidou, 2017; Tsakanikou *et al.*, 2021). Last but not least, a new generation of Greek students is being trained in Palaeolithic archaeology.

*Greek student training in Palaeolithic archaeology at Acheulean Lisvori-Rodafnidia, Lesbos.*

Since 2015 I have headed excavations in the Panthera Cave on Kythros, a small, barren island in the central Ionian Sea. Our work on Kythros is yielding a rich and diverse Middle Palaeolithic record (Galanidou, 2018) and is tied to our long-term research in the Inner Ionian Archipelago and on Lefkas (Galanidou, 2015; Galanidou *et al.*, 2016). This research has a strong regional perspective and covers the coast, the islands and the seabed. Beyond the Panthera Cave, we are studying Middle Palaeolithic sea crossing, or mere swimming, in a closed and well-protected sea where the destination, the next piece of dry land, was visible and required one to cover relatively short distances by sea.

Mapping the seabed is part of a new research direction I have taken lately, due to my interest in the islands' early record (Sakellariou and Galanidou, 2016; Galanidou and Bailey, 2020). I want to understand how Pleistocene submerged landscapes changed as the sea level changed, when terrestrial bridges were opened, creating new conditions for Pleistocene populations to settle and migrate. This field is tremendously interesting (Bailey *et al.*, 2020; Galanidou *et al.*, 2020). During the glacial periods, many islands were joined to each other and to the mainland, and the islands of the Ionian Sea and the Eastern Aegean were occasionally joined to mainland Greece and the Asiatic coast respectively. Imagine the archipelagos of the East Aegean and that of the Central Ionian Sea as being like a team of swimmers holding hands underwater and all we can see today are their heads above the surface.

*Panthera Cave teamwork, on the rocky shore of Kythros islet, Ionian Sea.*

### What project or publication are you most proud of?

Palaeolithic Lesbos is my pride, not only for the stunning Large Cutting Tool collection that multiplies every field season and could illustrate any textbook on Acheulean technology, but also for its potential to make the Aegean Region visible in the Eurasian Lower Palaeolithic narrative. I am also proud of my *Journal of Anthropological Archaeology* paper on forager use of cave space, which is still widely cited despite being twenty years old (Galanidou, 2000).

### *What do you think has been the most revolutionary discovery in your field over the last 5 years?*

The concept of Continental Shelf Prehistoric Research, bringing the seabed into focus. It holds the promise to radically change the ways we approach hominin dispersals, not merely by its potential to increase the sample of sites and finds but also by its call to make truly interdisciplinary contributions and breathe fresh air into palaeoanthropology, very much like the Neanderthal Genome Project did a decade ago.

*The first in situ hand-axe coming out of the Lisvori-Rodafnidia trench, August 2014.*

### *What has been your favourite memory from the field?*

The first handaxe that came out of Lisvori-Rodafnidia trench VI and was found in situ. I remember my heart was beating like a drum; something like the thrill of one's first kiss.

### *What would you be if you were not an archaeologist?*

A mathematician to engage with problem-solving or a politician to engage with building a better future.

*Lisvori-Rodafnidia, Lesbos, August 2013.*

***References:***


Bailey G. N., Galanidou N., Peeters H., Jöns H. and Mennenga, M. (2020). *The Archaeology of Europe's Drowned Landscapes* (Coastal Research Library Volume 35), Springer. Available online at: https://link.springer.com/book/10.1007%2F978-3-030-37367-2

Galanidou, N. (1997). 'Home is Where the Hearth is': The Spatial Organistion of the Upper Palaeolithic Rockshelter Occupations at Klithi and Kastritsa in Northwest Greece. BAR International Series 987.

Galanidou, N. (2000). Patterns in Caves: Foragers, Horticulturalists, and the Use of Space. *Journal of Anthropological Archaeology* 19(3), 243-275. DOI: 10.1006/jaar.1999.0362.

Galanidou, N. (2015). Seascape Survey on the Inner Ionian Sea Archipelago. In: Carver M, Gaydarska B, Monton S (eds), *Field archaeology from around the world. Ideas and approaches.* Springer, Cham, pp. 101–106

Galanidou, N. (2018). Parting the waters. Middle Palaeolithic archaeology in the central Ionian Sea. *Journal of Greek Archaeology* 3, 1–22.

Galanidou, N., Papoulia, C. and Iliopoulos, G. (2016). The Palaeolithic settlement of Lefkas Archaeological evidence in a palaeogeographic context. *Journal of Greek Archaeology* 1, 1-33.

Galanidou, N., Athanassas, C., Cole, J., Iliopoulos, G., Katerinopoulos, A., Magganas, A., and McNabb, J. (2016). The Acheulian site at Rodafnidia, Lisvori, on Lesbos, Greece: 2010–2012. In: Harvati, K., and Roksandic, M. (eds.), *Paleoanthropology of the Balkans and Anatolia*, Springer, Dordrecht, pp. 119–138.

Galanidou, N. and Bailey G.N. (2020). The Mediterranean and the Black Sea. In: Bailey G.N., Galanidou N., Peeters H., Jöns H., Mennenga, M. (eds.), *The Archaeology of Europe's Drowned Landscapes* (Coastal Research Library Volume 35), Springer, Dordrecht, pp. 309-319.

Galanidou N., Dellaporta K. and Sakellariou D. (2020). Greece: Unstable Landscapes and Underwater Archaeology, In: Bailey G.N., Galanidou N., Peeters H., Jöns H., Mennenga, M. (eds.), *The Archaeology of Europe's Drowned Landscapes* (Coastal Research Library Volume 35), Springer, Dordrecht pp. 371-392.

Sakellariou, D. and Galanidou, N. (2016). Pleistocene submerged landscapes and Palaeolithic archaeology in the tectonically active Aegean region. In Harff, J., Bailey, G. and Lüth, F. (eds). Geology and Archaeology: Submerged Landscapes of the Continental Shelf. *Geological Society, London, Special Publications*, 411, pp. 145–178 (first published online July 6, 2015, http://dx.doi.org/10.1144/SP411.9)

Sakellariou, D. and Galanidou, N. (2017). Aegean Pleistocene landscapes above and below sea-level: palaeogeographic reconstruction and hominin dispersals. In: Bailey G.N., Harf J., Sakellariou D. (eds.) *Under the sea: archaeology and palaeolandscapes of the continental shelf.* Springer, Berlin, pp. 335–336.

Tsakanikou, P., Galanidou, N. and Sakellariou D. (2021). Lower Palaeolithic archaeology and submerged landscapes in Greece: The current state of the art. *Quaternary International*, 584, pp. 171-181 (first published online June 3, 2020, https://doi.org/10.1016/j.quaint.2020.05.025)

# Part 2: Palaeoanthropology and biological anthropology

## Professor Rainer Grün[7]

Professor Rainer Grün is a world-renowned geochronologist, and Professor of archaeochemistry at Griffith University. He was the founding Director of the Australian Research Centre for Human Evolution and is an acknowledged leader in the field of electron spin resonance (ESR) and uranium-series dating (USR). His work has contributed immensely to our understanding of the timing of human evolution, for example the new, surprisingly young, dates from Broken Hill skull from Zambia (Grün *et al.*, 2020), which revealed that Africa and Eurasia were inhabited by a whole range of hominin species just a few hundred thousand years ago.

### *What are your research interests and area of expertise?*

I'm a geochronologist and the main thrust of my research relates to further developing dating techniques, in particular U-series and ESR dating (Grün, 2020). This requires expertise in geochemistry for U-series and radiation physics for ESR. In recent years, I mainly focussed on developing virtually non-destructive dating, which is essential for the analysis of ancient human remains (Grün, 2006). I've also ventured out to work on the use Sr isotopes for the reconstruction of human migrations and provided a Sr isotope map for France (Willmes *et al.*, 2018). Another sidetrack is working on palaeothermometry, which is a tool that can assess how fast valley erode and estimate the denudation rates of mountain ranges.

### *What originally drew you towards geochronology and human evolution?*

I was lucky to spend my first postdoctoral fellowship at McMaster University with Prof Henry Schwarcz. He got me interested in archaeological applications. In December 1986, I attended the International Colloquium L'Homme de Neanderthal in Liege, where I met Chris Stringer with whom I struck a lifelong friendship. He really got me into palaeoanthropological dating applications. Jean-Jacques Hublin attended the same meeting and our collaborations started at the same time.

### *What was your PhD topic? How did you find your PhD experience?*

My PhD was entitled 'Contributions to ESR dating'. It was pretty much a random selection of experiments on the underlying principles of ESR dating and applications on whatever samples I could get hold of. As an undergraduate student I was not very focussed on studying. I held several pinball machine records in the pubs around the Department of Geology in Köln. I thought my fate was becoming a taxi driver. However, once I discovered research during my thesis work, I got hooked. I worked pretty much around the clock and did my PhD in less than two and a half years.

[7] Australian Research Centre for Human Evolution, Griffith University; rainer.grun@griffith.edu.au

*Rainer meeting with Martin Aitken during Sr isotope fieldwork in France.*

### *After your PhD, what positions have you held and where?*

I was a postdoc fellow at McMaster University from mid 1985. In 1987, Ann Wintle asked me whether I was interested to take over her thermoluminescence lab at Cambridge. In the late 80s I met John Chappell who was the professor for Biogeography and Geomorphology at the ANU in Canberra. He encouraged me to apply for the directorship of the radiocarbon dating laboratory, which I commenced in early 1992. I stayed at the ANU for the next 23 years until I was offered to set up what later became the Australian Research Centre for Human Evolution at Griffith University in Brisbane.

### *What current projects are you working on? Where do you hope these go in the future?*

I am close to retirement. I have several projects that I pursue, in particular to understand the diffusion processes of uranium into bones and teeth and to see whether there are uranium concentration dependent correlations with the alpha efficiency in tooth enamel. I also continue to collaborate with a wide network of researchers on the direct dating of human remains. This work is presently divided up between Mathieu Duval who does the ESR analysis and I carrying out the U-series work.

### *What project or publication or discovery are you most proud of?*

They relate to original ideas. Truthfully, one does not have many of those in one's career. Two of the publications will be very obscure for the readers of this volume and they relate to the possibility of dating mollusc shells without the need to measure the external dose rate (Debuyst *et al.*, 1984; Grün, 1985), the other one is on the palaeothermometry of the Eldzurtinsky Granite (Grün *et al.*, 1999). The most important idea relating to palaeoanthropology was combining U-series and ESR systematics for the simultaneous modelling of the age of a tooth as well as the uranium uptake history of its various dental tissues (Grün *et al.*, 1988; Grün, 2000). This publication is the underpinning of all dating applications of teeth using these methods (Shao *et al.*, 2015).

*ESR team at Cambridge: John Chappell, Yuske Shimoyama, Ed Rhodes, Jack Rink, Henry Schwarcz.*

### *What do you think is the most revolutionary discovery in human evolution research over the last 5 years?*

Of course, the discoveries of several new human species such as *Homo naledi* and *Homo luzonensis* as well as the Denisovans. In particular, DNA work has brought a completely new dimension to our understanding of the complexity of human evolution: the interbreeding of what was thought to be different species, and the time depth of the occurrence of the various species.

### *What is the best thing about your job and what is one thing you would change if you could?*

I had a very satisfactory career and don't think I'd change a thing, mistakes, warts and all. Considering my somewhat forceful driving style, I don't think I would have made a good taxi driver.

*Rainer measuring samples from the Broken Hill skull with Chris Stringer (Grün et al. 2020). Photo credit: Katherine Griffiths.*

*Fieldwork in Namibia (2007).*

***References:***


Debuyst, R., Dejehet, F., Grün, R., Apers, D. and De Cannière, P. (1984). Possibility of ESR-dating without determination of the annual dose. *Journal of Radioanalytical and Nuclear Chemistry: Letters* 56(6), 399-410

Grün, R. (1985). ESR-dating without determination of annual dose: a first application on dating mollusc shells. *ESR-Dating and Dosimetry (IONICS, Tokyo, 1985)*, 115-123.

Grün, R. (2000). An alternative model for open system U-series/ESR age calculations: (closed system U-series)-EST, CSUS-ESR. *Ancient TL* 18(1), 1-4.

Grün, R. (2006). Direct dating of human fossils. *American Journal of Physical Anthropology* 131(S43), 2-48. DOI: 10.1002/ajpa.20516

Grün, R. (2020). A very personal, 35 years long journey in ESR dating. Quaternary International 556, 20-37. DOI: 10.1016/j.quint.2018.11.038

Grün, R., Schwarcz, H.P. and Chadam, J. (1988). ESR dating of tooth enamel: Coupled correction for U-uptake and U-series disequilibrium. *International Journal of Radiation Applications and Instrumentation. Part D. Nuclear Tracks and Radiation Measurements* 14(1-2), 273-241. DOI:10.106/1359-0189(88)90071-4

Grün, R., Tani, A., Gurbanov, A., Koshchug, D., Williams, I. and Braun, J. (1999). A new method for the estimation of cooling and denudation rates using paramagnetic centers in quartz: A case study on the Eldzhurtinskiy Granite, Caucasus. *Journal of Geophysical Research: Papers on Geodesy and Gravity Techonophysics* 10(B8), 17531-17549. DOI: 10.1029/1999JB900173

Grün, R. Pike, A., McDermott, F., Eggins, S., Mortimer, G., Aubert, M., Kinsley, L., Joannes-Boyau, R., Rumsey, M., Denys, C., Brink, J., Clark, T. and Stringer, C. (2020). Dating the skull from Broken Hill, Zambia, and its position in human evolution. *Nature* 580, 372-375. DOI: 0.1038/s41586-020-2165-4

Shao, Q., Chadam, J., Grün, R., Falguères, C., Dolo, J-M. and Bahain, J-J. (2015). The mathematical basis for the US-ESR dating method. *Quaternary Geochronology* 30, 1-8. DOI: 10.1016/j.quageo.2015.07.002

Wilmes, M., Bataille, C.P., James, H.F., Moffat, I., McMorrow, L., Kinsley, L., Armstrong, R.A., Eggins, S. and Grün, R. (2018). Mapping of bioavailable strontium isotope ratios in France for archaeological provenance studies. *Applied Geochemistry* 90, 75-86. DOI: 10.1016/j.apgeochem.2017.12.025

## Dr Briana Pobiner[8]

Dr Briana Pobiner is a palaeoanthropologist and the Education and Outreach lead at the Smithsonian Institution's Human Origins Program. Briana joined the Smithsonian National Museum of Natural History in 2005 to help establish the Hall of Human Origins, where her role now includes the management of public programs, website content, social media, and exhibition volunteer training. Briana is also an Associate Research Professor of Anthropology in the Centre for the Advanced Study of Human Paleobiology at the George Washington University. Her research is focussed on the evolution of the human diet, though she has studied topics as diverse as human cannibalism and chimpanzee carnivory.

### *What are your research interests and your particular area of expertise?*

My research focuses on the evolution of human diet, particularly surrounding meat-eating in our evolutionary history. I'm most interested in the earlier part of this dietary shift, between about 3 and 1 million years ago, documenting and trying to understand how meat became a more important part of ancient human diets. I do this by studying fossil animal bones that have butchery marks (cut marks from slicing off meat and percussion marks from breaking bones to access marrow) left by ancient humans. I also study the chewing patterns left by non-human predators on the bones they ate, so I can understand what parts of prey animals both humans and other predators were getting access to in the past.

### *What originally drew you towards human evolution?*

I started my undergraduate degree at Bryn Mawr without a strong interest in science – I was planning to be an English major and possibly pursue a career in creative writing. As I was looking for a fourth class to round out my first semester, my advisor, a dean who was a former Anthropology professor, suggested I take an Anthropology class called 'Introduction to Physical Anthropology and Archaeology'. I had never heard of anthropology, but it sounded interesting, so I signed up. I really enjoyed it, and then during the next semester I took a Primate Evolution and Behavior class with the professor who would become my main advisor (Dr. Janet Monge). I spent that summer doing a paleontology internship at the American Museum of Natural History, including fieldwork collecting invertebrate fossils, and loved it – but I was still really drawn to our own evolutionary history. I ended up creating an independent major called 'Evolutionary Studies' which included classes in biological anthropology, biology, ecology, geology, and paleontology. After my third year of college, I attended a field school in South Africa through the University of the Witwatersrand run by Lee Berger, got fully hooked on paleoanthropology, and I never looked back!

[8] Smithsonian Institution Human Origins Program; pobinerb@si.edu

## *What was your PhD topic? How did you find your PhD experience?*

My PhD research included two separate components. The first one was collecting and studying bones chewed on by free-ranging carnivores at Ol Pejeta Conservancy in Kenya, to try to document predator taxon-specific tooth marks and chewing damage patterns, with an aim to eventually look for similar patterns in fossil assemblages. The second one was studying collections of fossils with butchery marks and carnivore tooth marks from Koobi Fora, Kenya and Olduvai Gorge, Tanzania – I did multiple years of excavations with larger teams working at both locations while I was a PhD student. I really loved doing both aspects of my PhD research. I remember my PhD advisor – Rob Blumenschine – telling me as I finished classes and started doing my PhD research full time that I would never have the same kind of opportunity again, to be solely focused on research (with no other professional obligations) – and to make sure to savour and enjoy it!

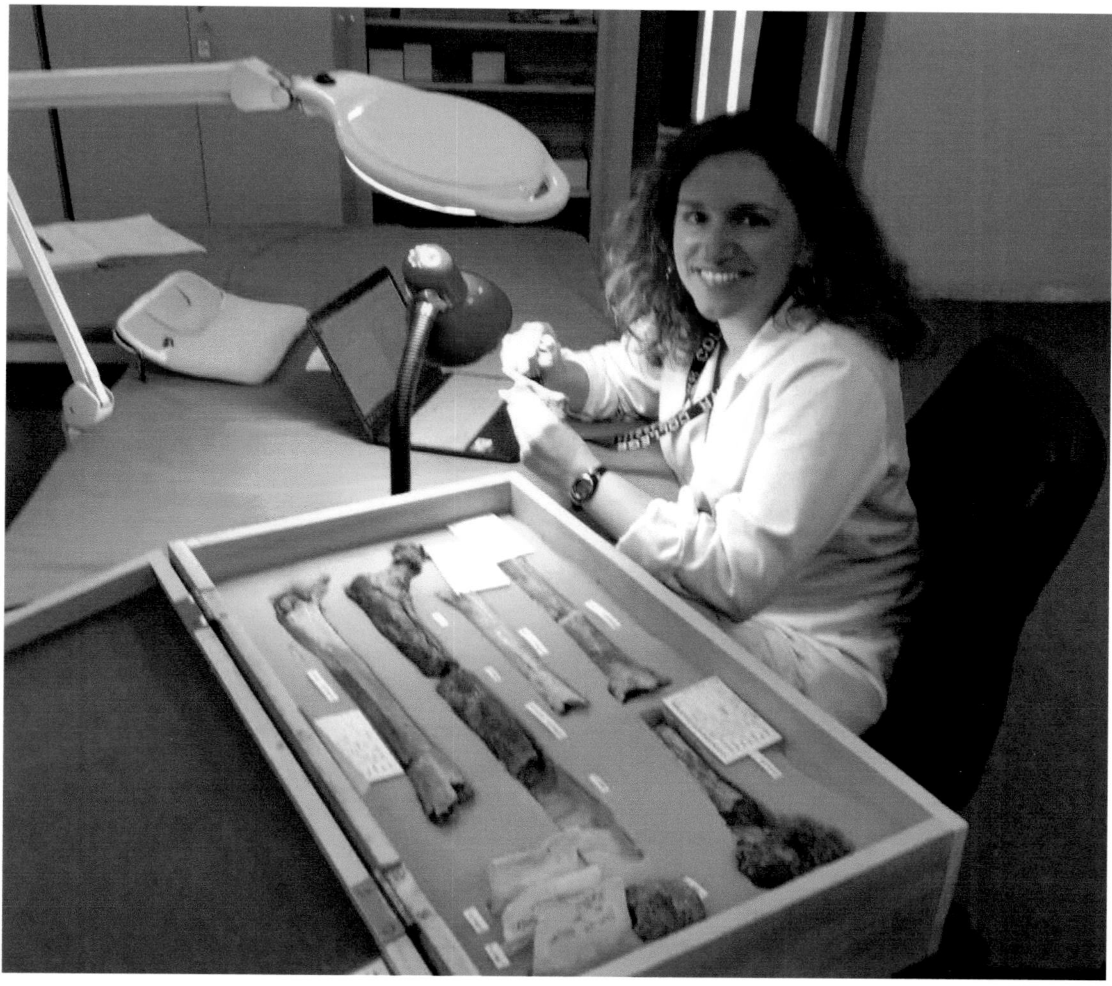

*Studying hominin fossils at the National Museums of Kenya.*

## *After your PhD, what positions have you held and where?*

I was invited by Rick Potts, the Director of the Human Origins Program at the Smithsonian's National Museum of Natural History, to become a predoctoral fellow even before my PhD was finished – and I have been working there ever since! My predoc fellowship became a postdoc fellowship once I finished my PhD, and during my postdoc I spent a lot of my time as a member of the core team that developed the permanent Hall of Human Origins at the National Museum of Natural History. I got very interested in public engagement with science during the exhibition development process, and after my postdoc ended, I got a unique permanent position at the museum that includes both research and public engagement. I lead the education and outreach efforts of the Human Origins Program, which includes content development for our website, managing social media (Facebook and Twitter) accounts, training and helping to manage volunteers in the exhibit, facilitating public programs, and participating in museum-wide education and outreach teams. I've also started an additional research program in the teaching and learning of evolution in high school biology classrooms! In addition to all that, I am an Associate Research Professor in the Center for the Advanced Study of Human Paleobiology at the George Washington University, where I regularly teach classes in both zooarchaeology and science communication.

*Briana talking to students during a Smithsonian outreach programme.*

### *What current projects are you working on? Where do you hope these go in the future?*

I have several current projects! I'm not technically finished with my postdoc project, which is collecting data on the fossil animal bones excavated from the oldest layers of the excavations at Olorgesailie, Kenya. I directed the field camp there for several years under the leadership of Rick Potts (who is also my supervisor), until my son was born – over 9 years ago now – and I decided that three months in the field every summer was too long to be away from my family. (You can read about how I attempt to balance motherhood with a career in paleoanthropology, including bringing my son to Kenya with me to do research in 2018, in a blog post for the Smithsonian https://www.smithsonianmag.com/blogs/national-museum-of-natural-history/2019/05/12/how-balance-motherhood-and-stem-career/).

Hopefully the Olorgesailie research will be all wrapped up in the next few years and will result in a comprehensive monograph.

I'm leading a long-term taphonomy and ecology research project on modern bones at Ol Pejeta Conservancy that I began during my PhD research there (BONES: Bones of Ol Pejeta, Neotaphonomic and Ecological Survey) together with wonderful research collaborators Fire Kovarovic, Kari Lintulaakso, and Ogeti Mwebi. I even did one field season there when I was very pregnant with my son! More recently, I've also been invited to work with a research team led by Claire Terhune and Sabrina Curran restudying previously excavated Pleistocene fossils from the Oltet River Valley in Romania to look for possible butchery marks, and I'm working with Michael Pante on some potential early evidence for human cannibalism (in the form of butchery marks on a Pleistocene early human fossil from Kenya).

Since I can't currently travel for research due to COVID-19, I'm also making progress on a few data-based research projects with various other collaborators.

*Briana doing fieldwork at Ol Pejeta Conservancy, Kenya, in 2011 while 7 months pregnant.*

### *What is your favourite memory from the field?*

Wow, I have so many, because I've spent so much time in the field and had so many field adventures that it's hard to choose just one! Here's a good one: I pride myself on being a skilled field driver who can manoeuvre a vehicle out of almost any situation – although my old Kenyan Land Cruiser was not always in the best shape (I can just hear some of my colleagues who have driven in it snickering right now). Once, when I drove it from Kenya to Tanzania to participate in fieldwork at a Pleistocene human footprints site called Engare Sero, several things went very awry while I was driving several of my collaborators out to the field site in a place where the word 'road' is a loose interpretation. One was that the rear axle – well, one of the half shafts – snapped, so I had to drive in low range, and whenever we leaned too far to that side one of my colleagues had to stick himself out the window to bang the half shaft back in with a shovel so it wouldn't slide out entirely. Then, my brakes failed – but I decided to see how long I could drive without any brakes and not have the passengers figure out what was happening, just by downshifting to slow down. After about 20 kilometres, I came to a part of the road where the lack of brakes became very apparent, and when they asked me why the vehicle was rolling backwards back down the hill we were on, it took me a while to stop laughing so I could explain the situation!

*Briana's Kenyan Land Cruiser temporarily stuck in a hole at Ol Pejeta Conservancy, Kenya.*

### *What project or publication or discovery are you most proud of?*

I'm really proud of all of my research projects and publications, but I'm currently still basking in the glow of having a sole author review paper 'The zooarchaeology and paleoecology of early hominin scavenging' published in *Evolutionary Anthropology* (Pobiner, 2020). Ever since graduate school I've really enjoyed the mix of comprehensiveness and accessibility of the review papers in that journal, and it feels like a real accomplishment to have now written one myself. Since I'm not in a traditional tenure-track faculty position I don't directly supervise graduate students, but I regularly take on undergraduate and graduate students as interns. I am always so proud to see them going on to do exciting and fulfilling careers, either in paleoanthropology, science education, or whatever makes them happy!

### *If you were not a palaeoanthropologist, what would you be?*

I loved my field ecology class in college and so much enjoy the field seasons at Ol Pejeta Conservancy – I think I'd probably be a field biologist. Or maybe a teacher, since I also get really jazzed seeing other people's faces light up when they understand something for the first time or make a new connection.

*Briana standing next to a replica of the 'Lucy' skeleton in the Smithsonian's Hall of Human Origins.*

***If you had a time machine, how far would you ask to go back, where would you go, and what would you want to see?***

I love this question! I was a member of the team who studied the butchery-marked fossils from Kanjera South, Kenya, where the earliest evidence of repeatedly transporting animal carcasses to the same location for butchery has been documented at about 2 million years ago. I'd want to watch those early humans, or really any early humans from around this time period, when this important behavioral shift happened. How were they cooperating? How did they communicate? How did they decide which animals to transport and butcher, and what parts to process for meat and marrow? How important were animal foods to them within the rest of their dietary choices – and what else were they eating? And who decided that already dead animals might make a good meal – who were those early scavengers? Basically, I'd want to see what was on the paleo-menu, and how the early humans made those decisions!

***References:***

Pobiner, B. L. (2020). The zooarchaeology and paleoecology of early hominin savenging. *Evolutionary Anthropology: Issues, News and Reviews* 29(2), 68-82. DOI: 10.1002/evan.21824

## Dr Mirriam Tawane[9]

Dr Mirriam Tawane is the Curator of the Plio-Pleistocene collection at the DITSONG: National Museum of Natural History in South Africa. She was awarded her Bachelor of Science degree, Master of Science degree and PhD in Paleoanthropology all from the University of Witwatersrand, becoming the first black South African female to qualify as a palaeoanthropologist in 2012. Mirriam has been involved in many public engagement initiatives over the years, as she is passionate about her country's heritage and public engagement in human evolution. She has participated in several community projects to teach the general public about the palaeosciences in South Africa, the majority of which are carried out in the language choice of the audience

### *What are your research interests and particular area of expertise?*

I spent most of my studies focusing on dental morphology. I have been collaborating on research topics aligned with dentition of hominins. I am also doing a lot of outreach focusing on teaching human evolution at schools. It is a project that has been ongoing for some time, and with it mushrooms projects that we implement to mitigate the situations we come across. These could be lack of teaching materials, or teachers and scholars needing assistance regarding the subject.

### *What originally drew you towards palaeoanthropology?*

While in high school, I had no interest in the subject. To be honest, I was not even aware of such a career choice. It was only when I did a Palaeontology course taught under Zoology in my third year that I became aware of such an option. I grew up in one of the villages in Taung, about 25 kilometres from Buxton Village. Buxton village is a village where the Taung skull was discovered. During one of Palaeontology lectures, I was introduced to the scientific information about the Taung skull, its discovery, and the role and significance its discovery has played in all that we know regarding the origin of Mankind. Having only known what I could label 'the village gossip' regarding the skull and realising the lack of participation (or participation in limited numbers) of people of colour in the field, I was motivated to pursue the course all the way to PhD level.

### *What was your PhD topic? How did you find your PhD experience?*

My PhD topic was 'Dental size and frequency of anomalies in the teeth of a small-bodied population of Mid-Late Holocene Micronesians, Micronesia'. I worked on dental remains of specimens Prof Lee Berger discovered in the Palau Islands. The specimens recorded very large teeth compared to their short stature and small brain. These could be attributed to diet and possibly hereditary features.

My PhD experience was exciting and scary at the very same time. As most of us will experience a lot in our lifetime, there are usually those unspoken criticism that one is subjected to, they end up making you doubt yourself, your capabilities. Having said that I need to admit that I was my biggest critic during those years. Everything I did, I had to redo, double check; just to make sure that it is the best I could do

---

[9] DITSONG: National Museum of Natural History, South Africa; tawane@ditsong.org.za

at that particular moment. My biggest challenge was to tell myself 'Relax, you have made it this far. You are going to make it'.

## *After your PhD, what positions have you held and where?*

I was a postdoctoral fellow at the University of the Witwatersrand for few years. I worked on several projects. One was looking at some of the remains from Sterkfontein caves. I discovered the hominin first rib of *Australopithecus africanus* from Member 4. Upon analysis, we determined that it falls closest to the small-bodied Australopithecines (AL 288-1 and MH1).

I also worked on a stakeholder analysis of all the stakeholders involved at the Taung Skull Fossil Site. This was to determine the status of development of the site, and how all the stakeholders relate and work together towards achieving this. I also participated in outreach projects to deliver the much-needed information about the site, the skull to the communities within the vicinity of the site. These targeted the scholars in the form of workshops at schools as well as the general public in the form of a Heritage Day celebration hosted on the 24 September. The 24th September is National Heritage Day in South Africa.

*Human evolution workshops at schools in Taung.*

### *What current projects are you working on? Where do you hope these go in the future?*

I am currently working on a national project to get all museums in the country curating Humanities-related collections working together for the benefits of the heritage objects. This initiative is to address certain objectives necessary for the preservation and safeguarding of the collections. Collection housing institutions tend to be fragmented and isolated. What is happening on the one hand is usually not known by the other hand. Implementing best practices and common standards with regards to a museum environment, focusing mainly on humanities collections is also prioritised. The aim is to standardise collection care, although there will be exceptions here and there. I am hopeful that this project will bring curators curating humanities-related collections across South Africa closer together, and that there will be collaboration between museums and ultimately skills transfer, as well as the achievement of the main goal of this initiative: to safeguard and preserve the heritage collections in the country.

### *What project or publication or discovery are you most proud of?*

While working at Taung, together with a colleague at Wits, we sourced funding to present human evolution workshops at schools. The aim of these workshops was to present the Palaeosciences as a career choice and to bridge a gap that existed regarding evolution and related subjects that exists among communities living a stone throw away from the site. We compiled worksheets and human evolution teaching packs to donate to schools.

Evolution is a complex subject to learn. We introduced a form of 'edutainment' to the project. Participating schools were tasked to create songs using the site, it's discoveries as a focus point. That was very successful, as we ended up recording 9 songs that are both educational and very entertaining. The scholars took the challenge very seriously.

### *What do you think is the most revolutionary discovery in human evolution research over the last 5 years?*

I might be biased and focus on those that are perhaps close to home. I will have to mention the discovery of *Homo naledi* in 2015. Little foot might have been discovered in 1994, but it was in 2017 that it was unveiled for the world to see. The near completeness of the specimen is remarkable. I believe those discoveries and the continuous research on the specimens will add on to the knowledge we have regarding human evolution. One also needs to acknowledge that technological advancement in the field is allowing for ground-breaking research to be undertaken on materials that are somewhat difficult to study. I work in a museum, where we curate thousands of pieces of specimens, or rather heritage objects as they are mostly called in the museum environment, collected a long time ago. These pieces should be regarded as active chess pieces with the potential to contribute immensely to the active discussions and discoveries currently taking place.

### *What is the best thing about your job and what is one thing you would change if you could?*

The best thing about my job is teaching. I do lectures and tours of the collections to scholars and the general public. When you present a tour of the collection to scholars with little knowledge about evolution, and you start to observe them grasping the concept and the confusion slowly disappears from their faces, that comes with some form of contentment. The one thing I would change, though not entirely do away with, is the amount of administration that one has to do. Yes, administration is a significant role of any position, and it should be taken very seriously but I would streamline reporting in a way that few reports need to be put together and they would be suitable to be submitted to different departments or line managers.

*Human evolution workshops at schools in Taung.*

## Dr Trish Biers[10]

Dr Trish Biers is the Curator of the Duckworth Laboratory in the Leverhulme Centre for Human Evolutionary Studies at the University of Cambridge. As well as curating and managing the Duckworth's human remains collections, Trish teaches about treatment of the dead, ethics, and decolonisation for the Department of Archaeology and runs courses at the Institute for Continuing Education at Cambridge. Previously, she has held positions in curation, osteology and outreach at the Repatriation Osteology Laboratory in the Smithsonian Institution, Museum of Archaeology and Anthropology in Cambridge, and San Diego Museum of Man in California. She currently serves as Museum representative on the Board of Trustees for the British Association for Biological Anthropology and Osteoarchaeology and cofounded MorsMortisMuseum – a website dedicated to the role of human remains in museums.

### *What are your research interests and particular area of expertise?*

My research interests revolve around death and human remains, decolonising the dead, and ethical issues about displaying the dead in museums. I'm also interested in Andean archaeology, gender, cemetery and graveyard research, and folklore studies in witchcraft and magic and material culture. My areas of expertise are osteology and paleopathology, museum curation and conservation, scientific investigations of human tissues, ethics and repatriation.

### *What first inspired your interest in osteology and paleopathology?*

As a teenager I went to the San Diego Museum of Man (soon to be the Museum of Us) all the time. I was always interested in death, skeletal structures, mummified remains and burials, and forensics. At 19, I got an internship with the Physical Anthropology collections under Rose Tyson, a phenomenal osteologist and palaeopathologist, and she trained me. I got a job there at 21 and worked my way up from 'shop girl' to Associate Curator. I also volunteered with a forensic entomologist named Dave Faulkner who helped me develop my academic trajectory. At the museum, we hosted incredible scholars from all over the world including the late, great, Dr Don Ortner whose knowledge of pathology was remarkable – I learned so much. I worked on collections and exhibitions while doing my Undergraduate and Master's degrees, I was there for 11 years!

### *What was your PhD topic and who was your supervisor? What were the findings from your PhD?*

My PhD was titled, 'Investigating the Relationship between Labour and Gender, Material Culture, and Identity at an Inka Period Cemetery: a regional analysis of provincial burials from Lima, Peru.' It combined human skeletal data, burial deposition, and documentary sources to assess identity of artisans under Inka (AD1400-1532) provincial control. My supervisor was Dr Elizabeth DeMarrais and my advisor was Dr John Robb here at Cambridge. I found some really interesting patterns in burial style

[10] Duckworth Laboratory in the Leverhulme Centre for Human Evolutionary Studies at the University of Cambridge; tmb40@cam.ac.uk

(six types of mummy bundles), grave associations and gender, in particular those associated with older women, and women's labour under Inka rule. They were very skilled artisans!

*Trish with her colleague Bertha excavating a Peruvian mummy bundle in Lima (Photo credit: Elena Goycochea).*

### *What projects are you currently involved with? Where do you hope these will go in the future?*

I have a few things happening at the moment as well as just surviving this crazy pandemic! I'm working on a proteomics/genomics project with colleagues from Peru and the US that has had a successful pilot project so we will write a grant to do more in-depth osteobiographies. I've just had a book proposal accepted with my colleague about museums, heritage and death (Biers and Clancy, forthcoming 2022). I'm thrilled to be involved in the Legacies of Empire research group via the University of Cambridge Museums (UCM) and am excited to see how we can dismantle the colonial collecting practices of the past associated with Cambridge collections. And finally, I'm having fun with new research about witchcraft, human remains, and material culture.

### *What does your role as Curator of the Duckworth Laboratory at the University of Cambridge involve?*

Well, I research and conserve the collections and archive on a daily basis. This means I check environmental conditions across the Duckworth spaces, re-box and catalogue remains, photograph remains for the database, prepare remains for repatriation, do archival research, and facilitate researcher access to the collections. I build protective structures for more fragile collections and document conservation work. I'm trying to make the collections more accessible with up-to-date information to eventually be put online. I also consult about human remains collections with other institutions in Cambridge offering advice and collaborative strategies and participate in outreach initiatives to engage the public with the work we do (Biers and Harknett, 2015).

*Trish engaging with the public at the 'Plague Late' evening event at the Museum of Archaeology and Anthropology, Cambridge.*

**What type of research is done on the collections in the Duckworth Laboratory?**

We host all sorts of people in the Duckworth including visiting researchers, undergraduate and MPhil students, PhDs and post-docs, and individuals doing archival research. The scope is broad, from non-human primate anatomy to hominin development to dental morphology. It's mostly dissertations about human skeletal remains and a lot of researchers use micro-CT and various types of photogrammetry for 3D modelling. Applications for destructive sampling are on the rise as well.

**You have also worked at the Smithsonian Institute in their Repatriation Osteology Laboratory. Are there many differences between the UK and the USA in terms of protocols, ethics etc. when dealing with human remains?**

Yes and no. In regard to ethics, those are pretty standard across the biological anthropology and archaeology communities whether university/museum or commercial units, as professional organisations have codes of ethics that we are supposed to follow (for example, see British Association of Biological Anthropology and Osteoarchaeology, 2019). and these are very similar internationally, i.e., dignity and respect of human remains.

What's different is legislation, and the US has the Native American Graves Protection and Repatriation Act (NAGPRA). Enacted in 1990, Federal law provides for the repatriation and disposition of Native American/First Nations human remains, funerary objects, sacred objects, and objects of cultural patrimony. It is a robust programme and is now often used as a model for repatriation discourse and strategy globally. In the UK, repatriation is much less structured and there are national guidelines, but institutions have more options in compliance. This is changing with a national movement to 'provide new guidance for the UK museum sector on the restitution and repatriation of cultural objects'.

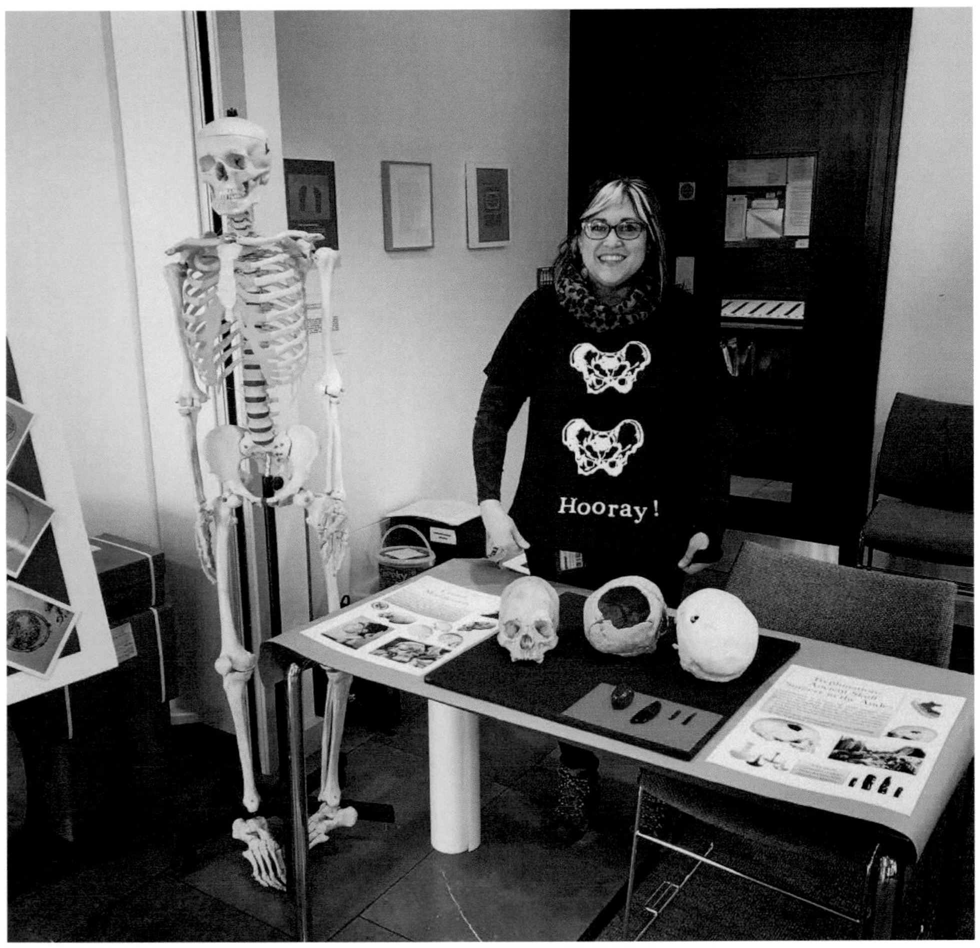


*Trish presenting on trephination and cranial modification with skeletal casts at the Cambridge Science Festival.*

### *What is the best thing about your job?*

I'm lucky that I get to teach on ethics, treatment of the dead, osteology, and repatriation for the Department of Archaeology. One of my favourite parts of the job is working with students and helping them with their projects. I'm thrilled to be advising a PhD student on her work with human remains abroad and we are having so much fun together (and being serious academics, of course). In August, I usually have several students helping me with re-boxing remains and it is enjoyable to see them get invested in the care of the collections. On the other hand, I also like disappearing into the collections and archives and having a quiet space for research and reflection.

*Trish, alongside colleague Sarah-Jane Harknett from Museum of Archaeology and Anthropology (MAA), presenting at the Death, Dying, and Disposal Conference in Bath 2019.*

### *What advice would you give to someone who is interested in a career working with human remains?*

Death is such a deep, philosophical entity. How you perceive death and a dead body, whether it's skeletonised, fleshed, ancient or new, can influence how you work with say, archaeological human remains, or rather how you perceive them. Are they biomatter? Are they ancestors? Who really cares if they are dead? It's all tied up into ideology, really, despite a background in the scientific method (it's fascinating to see how different the views are amongst my friends/colleagues). I personally think it is important to be mindful of death practices globally in the past and the present because there is SO much variation in how humans treat dead bodies both physically and spiritually (Biers, 2020). This can help you build your professional narrative during your skeletal biology and anatomy studies in addition to field and lab methods. If you are more 'museumy' in nature, then be prepared for the emotional and troubling information you can come across if you are working with collections that stem from colonial/imperial collecting practices. Be curious but be thoughtful!

### *References:*

Biers, T. and Clary, K (eds) (forthcoming May 2022). The Routledge Handbook of Museums, Heritage, and Death. Routledge: New York.

Biers, T. (2020). Rethinking Purpose, Protocol, and Popularity in Displaying the Dead in Museums. In: Squires K., Errickson D., Márquez-Grant N. (eds), *Ethical Approaches to Human Remains.* A Global Challenge in Bioarchaeology and Forensic Anthropology. Springer. pp 239-263.

British Association of Biological Anthropology and Osteoarchaeology (2019). BABAO Recommendations on the Ethical Issues Surrounding 2D and 3D Digital Imaging of Human Remains. Available online at: https://www.babao.org.uk/assets/Uploads/BABAO-Digital-imaging-code-2019.pdf

Biers, T. and Harknett, S. J. (2015). Separating Artefact from Fiction: using museum education and outreach to increase archaeology's relevance and impact in society. *Archaeological Review from Cambridge. Archaeology: Myths within and without.* Vol. 30. 2.

***Extra resources:***

- A Repatriation Resource: https://padlet.com/emmalmartin73/55eq3rdjdn7j Museum Ethnographers Group
- Native American Graves Protection and Repatriation Act: https://www.nps.gov/subjects/nagpra/index.htm

## Professor Tanya Smith[11]

Photo credit: Jeff Camden

Professor Tanya Smith is a human evolutionary biologist at the Australian Research Centre for Human Evolution (ARCHE) and the Griffith Centre for Social and Cultural Research (GCSCR) at Griffith University. Tanya, following a PhD in Anthropological Sciences at Stony Brook University, has held fellowships at the Radcliffe Institute for Advanced Study and the Max Planck Institute for Evolutionary Anthropology, in addition to a professorship at Harvard University. Her research at ARCHE and GCSCR focuses on primate dental growth, using tooth microstructure to resolve taxonomic, phylogenetic and developmental questions about great apes and humans, as demonstrated by her recent popular science book *The Tales Teeth Tell*. She has published in a number of high-impact journals and her work has been reported in The Conversation, The New York Times, National Geographic, Smithsonian, and Discovery magazines. She has also appeared on American, Australian, British, Canadian, French, Irish, German, New Zealand, and Singaporean broadcast media. www.drtanyamsmith.com

### *What inspired your interest in human evolution and specifically primate teeth?*

I was initially lit up by an introductory biological anthropology course I took with Robert (Bob) Anemone during my first semester at SUNY Geneseo in 1993. The field encompasses so many personal interests in natural history, skeletal biology, and human uniqueness. While majoring in biology, I took every one of Bob's bio anthro courses and participated in two field seasons in the Great Divide Basin of Wyoming — where we recovered Eocene mammalian fossils, including tiny primate teeth. During my senior year at Geneseo I began to read about how scholars were using biological rhythms in teeth to explore ancient human development, and using electron microscopy, I started my own search for these lines in the fossil teeth we found in Wyoming.

### *What types of information can we obtain about human evolution from teeth?*

Nearly everything you can think of: birth, growth rates, age, disease, evolutionary relationships, life history, diet, migration, climate, nursing behaviour, and even social status — humans have used tooth modification as a form of personal expression for thousands of years. As we discuss below, I had no trouble filling an entire book with diverse types of information on teeth (Smith, 2018)!

### *What was your PhD topic? How did you find your PhD experience as a young woman embarking on a career in academia?*

As someone who happily counts tiny time lines in a dimly lit microscope room, I am drawn to empirical research on things that can be quantified precisely. For my PhD I studied the development of primate teeth, testing hypotheses about biological rhythms and methods to characterize their growth, as well

[11] Australian Research Centre for Human Evolution and Griffith Centre for Social and Cultural Research, Griffith University; tanya.smith@griffith.edu.au

as exploring variation in chimpanzee molar enamel (Smith, 2004). I was fortunate to have a supportive advisor at Stony Brook University, Lawrence Martin, who also employed me to run his laboratory and study Miocene ape dental development. At the time I was aware that some other faculty in the department were less supportive of women, but the challenges of being part of a marginalized academic group didn't come into real focus until later in my career.

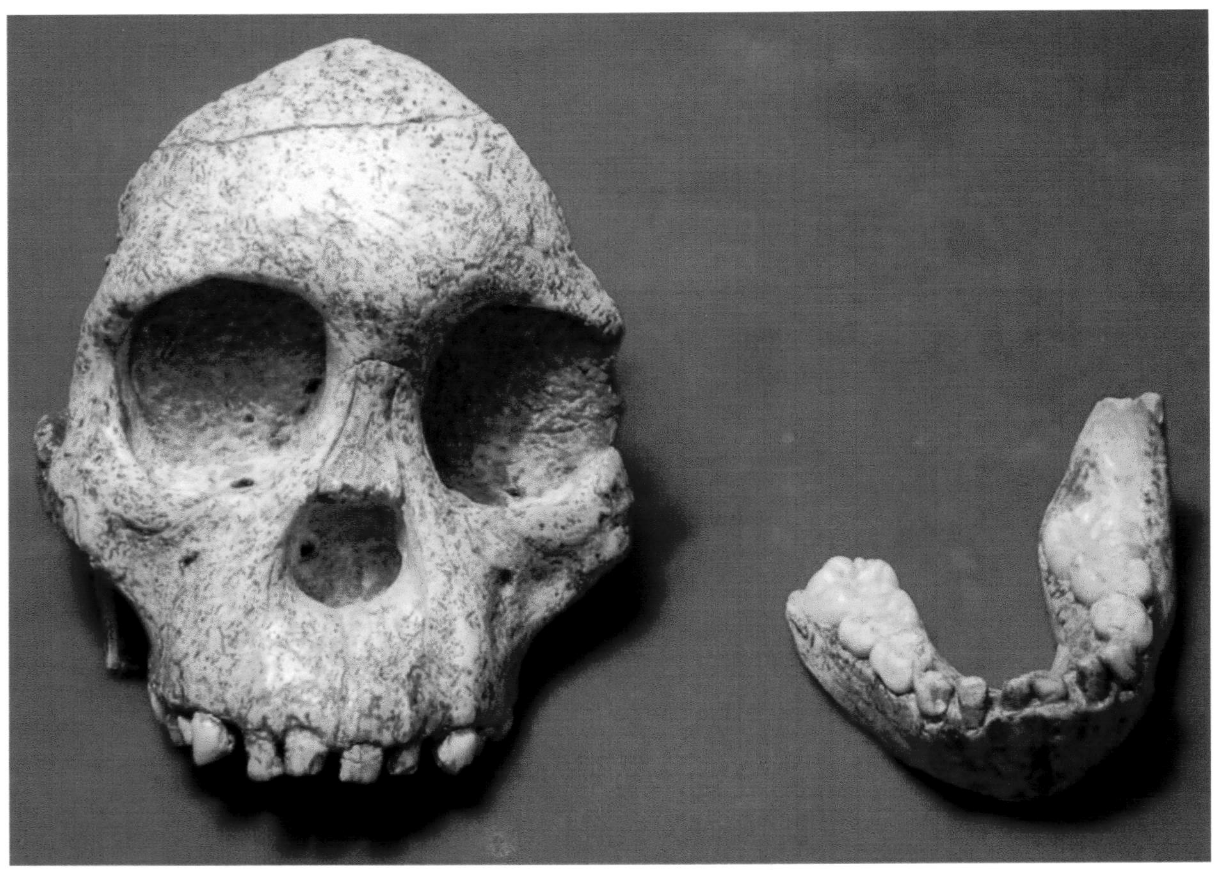

*Australopithecus africanus Taung child studied by Raymond Dart. The upper and lower jaws have a full set of baby teeth along with the permanent first molars. Fossil courtesy of the University of the Witwatersrand (Johannesburg); photo credit: Tanya M. Smith and MIT Press.*

***After your PhD, you've worked in a number of institutions in many countries all over the world. Do you think your development as a scientist has benefited from working in these diverse working environments?***

Unquestionably. I recently wrote an article for the US Association for Women in Science called Academics without Borders (Smith, 2020). Anthropologists emphasize cultural relativism — seeing differences without judgement — and there are fascinating contrasts in the way that scholars work in different parts of the world. My exposure to diverse academic cultures on several continents has helped me to work differently, including building long-lasting collaborations with some fascinating people. I find myself drawn to individuals who are not threatened by diversity or motivated by a sense that their work defines them, and really enjoy collaborations with international scholars whom I pepper with questions as I absorb their personal and professional perspectives. Novelty is perpetually engrossing to me.

*Tanya in Uganda in 2014 watching Bud, a male chimpanzee, whose dental development was detailed in Machanda et al. (2015). Photo credit: Zarin Machanda and MIT Press.*

***Recently, you published the popular book 'Tales Teeth Tell'. Tell us a little bit about the book and your experiences when writing it.***

It was actually a planned experiment in a sense; I wanted to both learn how to write a book and show that women in the middle of their careers can communicate about science to the public. The main work occurred during a sabbatical and overseas move, which slowed me down but kept me sane as I didn't have a lab to work in for more than a year. It's been wonderful to see other women in biological anthropology recently stepping into leadership in this way; the story of human evolution has classically been told by men nearing the end of their careers — yet biological anthropology is not just a boy's club!

### *What projects are you currently working on? What do you hope to find out?*

Even after twenty years of seriously thinking about how teeth grow, I am still on fire about them! My collaborators and I have recently been collecting isotopic data from primate teeth with an ion microprobe, and we are able to precisely document birth through distinct elemental shifts — which raises so many questions about what is going on inside mothers and infants during this profound transition, as well as what the cells are doing to create this permanent structural and chemical record. On a broader scale, I continue to noodle over how dental tissues may better resolve questions about the origin and evolution of humans. And I'm now starting an Australian Research Council Future Fellowship to investigate prehistoric human population growth by analysing ancient children before and after key cultural transitions.

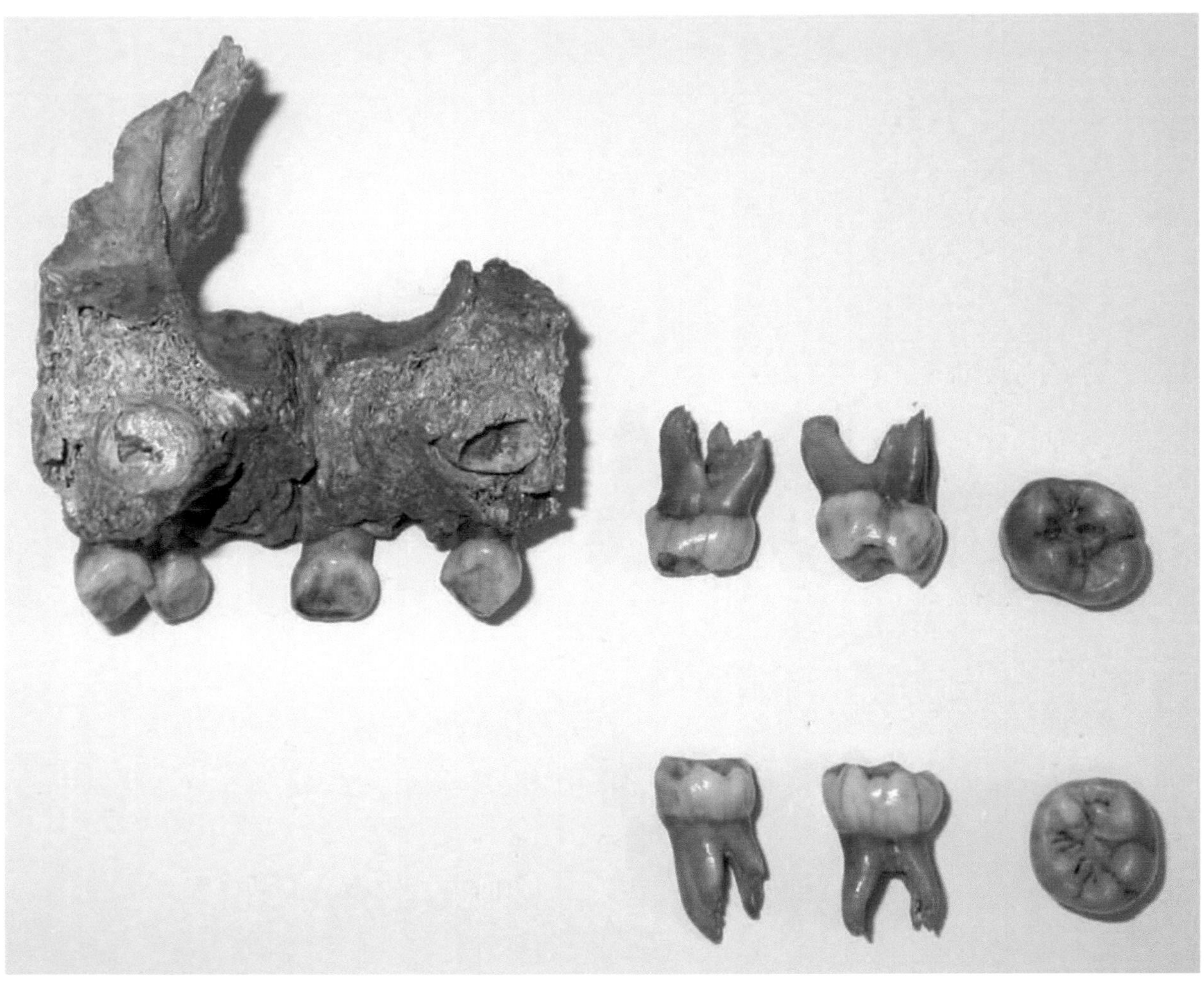

*3-year-old infant Neanderthal upper jaw and associated baby and permanent teeth, whose precise age was reported in Smith et al. (2010). Individuals from this Belgian site (commonly known as Engis) were the first fossil hominins ever discovered. Fossil courtesy of the University of Liège (Belgium); photo credit: Tanya M. Smith and MIT Press.*

### *What achievement are you most proud of?*

In 2018 I published a collaborative paper in *Science Advances* detailing how the teeth of Neanderthals can be used to reconstruct weekly records of ancient climate, nursing behaviour, and illness (Smith *et al.*, 2018). Leading an amazing group of anthropologists, earth scientists, and public health specialists to make these discoveries was one of my most satisfying accomplishments, and our team was one of three finalists for the Australian Museum's 2019 Eureka Prize for Excellence in Interdisciplinary Scientific Research. Another memorable achievement was co-hosting the Biological Anthropology Women's Mentoring Network's 10-year anniversary party at the American Association for Physical Anthropology meeting in April 2019. I co-founded BAWMN with some friends as women are underrepresented in palaeoanthropology and at the senior academic ranks, and I enjoy connecting with network members at annual meetings much more than partaking in the traditional conference activities (www.bawmn.org).

*Tanya enjoying sunrise on top of Big Red, the famous 40-metre-high sand dune at the edge of the Simpson Desert in Birdsville, Queensland. Photo credit: Tony Miscamble.*

### *What is the best thing about your job and what is one thing you would change if you could?*

I absolutely cherish having the professional freedom to pursue what I am curious about. It is an incredible privilege, and I try to encourage others to find confidence to do the same in whatever form that might take. One thing I would like to change about being an academic is the expectation that one should continuously train PhD students. I've had a chance to work with some amazing students and early career researchers, who are under incredible pressure in the current circumstances. Given long-term decreases in permanent academic jobs — I find it frustrating that some universities are graduating more and more PhDs each year.

### *If you were not a human evolutionary biologist, what would you be?*

I am enraptured with contemplative neuroscience, having had a meditation practice for over a decade, and I enjoy dipping into scholarly literature and popular science books on the topic. It's truly amazing to me that we can train our minds in ways that are analogous to conditioning our muscles at the gym. I've experienced the positive effects of mediation and spent time with some wonderful Buddhist nuns, and could easily imagine pursuing the academic study of meditation in a future lifetime.

### *References:*


Machanda, Z., Brazeau, N. F., Bernard, A. B., Donovan, R. M., Papakyrikos, A. M., Wrangham, R. and Smith, T. M. (2015). Dental eruption in East African wild chimpanzees. *Journal of Human Evolution* 82:137-144. DOI: 10.1016/j.jhevol.2015.02.010

Smith, T. M. (2004). Incremental Development of Primate Dental Enamel. PhD Dissertation. Stony Brook University. Available online at: http://paleoanthro.org/dissertations/download/

Smith, T. M. (2018). The Tales Teeth Tell: Development, Evolution, Behavior. MIT Press, Cambridge, MA, 296 pp.

Smith, T. M. (2020). Academics Without Borders. Association for Women in Science Magazine. July 2020, pp. 10-12.

Smith, T.M., Tafforeau, P., Reid, D. J., Pouech, J., Lazzari, V., Zermeno, J. P., Guatelli-Steinberg, D., Olejniczak, A. J., Hoffman, A., Radovčić, J., Masrour, M., Toussaint, M., Stringer, C. and Hublin, J-J. (2010). Dental evidence for ontogenetic differences between modern humans and Neanderthals. *Proceedings of the National Academy of Science of the United States of America* 107:20923-20928. DOI: 10.1073/pnas.1010906107

Smith, T. M., Austin, C., Green, D. R., Joannes-Boyau, R., Bailey, S., Dumitriu, D., Fallon, S., Grün, R., James, H. F., Moncel, M. H., Williams, I. S., Wood, R. and Arora, M. (2018). Wintertime stress, nursing and lead exposure in Neanderthal children. *Science Advances* 4(10), eaau9482. DOI: 10.1126/sciadv.aau9483

## Professor Rebecca Ackermann[12]

Professor Rebecca Ackermann is a biological anthropologist at the University of Cape Town (UCT). She was the founding Director of the Human Evolution Research Institute at UCT and is currently Deputy Director. She is also Deputy Dean of the Transformation in the Faculty of Science at UCT. Her research focusses on evolutionary process, and specifically how gene flow, drift and selection interaction to produce skeletal diversity through time, with a focus on human evolution. Rebecca is also engaged in discourse and policy development around sexism, racism and transformation of the discipline more generally.

### *What are your research interests and your particular area of expertise?*

I am a biological anthropologist who studies morphological – primarily skeletal – variation. I'm interested in knowing why we vary, why lineages diverge evolutionarily. This obviously involves getting a firm grasp of both within and between group variation, and I have looked at a lot of different organisms – from mice to gorillas to hominins – to do this. In particular, I've studied the relationship between patterns of variation and the evolutionary processes that produce them, i.e., the relative roles that selection, drift, and gene flow (hybridization) play in producing diversity (Ackermann and Cheverud, 2004; Ackermann *et al.*, 2006).

### *What originally drew you towards biological anthropology?*

I've always been interested in bones, since I was small. Someone recently reminded me of the story of how I was reading a magazine describing a child's struggle with osteogenesis imperfecta (brittle bones) when I was about 8 or 9, and decided I wanted to find a cure for it. Although I didn't go into that line of research, clearly I have remained hooked on bones! But more than that, I wanted to understand what makes us different and why. That included trying to understand everything from human variation (and race and racism; Fuentes *et al.*, 2019) all the way through to fossil hominin taxonomic diversity. I was very lucky to have great mentors when at The University of Chicago as an undergraduate (special shout out to Jane Buikstra), who really helped me to explore all of anthropology and come out with a more holistic approach to considering this question. That was the time when I decided biological anthropology was for me.

### *What was your PhD topic? How did you find your PhD experience?*

I studied facial variation in what were then the earliest known hominins – the australopiths. A large part of that research examined how our underlying assumptions about how and why hominin taxa vary shape our species-determinations. I grew up in the United States, and my PhD work was the first time I was privileged to travel to Africa, first to South Africa in 1995 for preliminary work, then to South Africa and Kenya in 1996/7 for data collection. Like most foreigners coming to these spaces for the first time, I found it incredible. To be able to see and touch the fossils for yourself, meet people such as Phillip Tobias, and experience a very different culture. 1995 was also when South Africa won the Rugby World Cup in their first participation post-apartheid, and that was an experience I will never forget. But my

[12] Human Evolution Research Institute, University of Cape Town; becky.ackermann@uct.ac.za

PhD experience was not all rosy. Multiple times during my academic training I experienced sexual harassment, and this forever shaped me. I changed universities because of it, avoided certain academics and curators, and ultimately modified my choices going forward. During that time and for many years after, I also felt the weight of being a woman and not being included or taken seriously in the discipline and was repeatedly bullied at conferences and in other academic spaces, even when I was supposed to be the authority. These experiences had a profound effect on me and on my choices going forward, and on my mentoring especially.

### *After your PhD, what positions have you held and where?*

I did a two-year postdoc at the same institution where I received my PhD – Washington University in St Louis – under the supervision of James Cheverud. Halfway into that, the advertisement for a Lectureship in the Archaeology Department at the University of Cape Town crossed my desk. I only realised later that the post was advertised with what we now call transformation goals in mind – i.e., to hire a black or female South African. I applied, and ultimately was offered the position. While in South Africa previously I had been to Cape Town and said I would move there in a heartbeat if I got a chance. So, I did. In 2000, I moved with myself, my husband, and our three old dogs, and have been here ever since. But the fact that there were no qualified South Africans to take up the position bothered me from day one, and I made a commitment to myself that one of my primary goals would be to make sure that next time there would be. I am now Professor and Deputy Dean for Transformation in the Faculty of Science, and my job is to continue the work of transforming our institution to one that reflects the demographic and cultural diversity of South Africa.

*Rebecca at home in Cape Town, with Table Mountain in the background.*

### *What current projects are you working on at UCT? Where do you hope these go in the future?*

I am involved in quite a lot of research projects, many of which are in collaboration with current or former students. In addition to my focus on skeletal morphology, I am involved in issues around decolonisation in palaeoanthropology (Athreya and Ackermann, 2020). Disparities in wealth, opportunities and privileges in the discipline have meant that the demographic of who gets to ask and answer research questions has historically been, and still is, skewed to the West. These disparities have grown out of colonial/patriarchal practices, and their correlates, racism and sexism. We're paying more attention to this globally, especially right now, but need to look at ourselves more critically, and especially how we as individuals and collectives continue to prop up these systems and impede the transformation of our discipline.

*Rebecca with her colleagues and her son in the field (left to right: Rebecca, Nomawethu Hlazo (PhD candidate), Dr. Lauren Schroeder, Dr. Job Kibii, her son Zane, Dr. Robyn Pickering) at the Cradle of Humankind in 2019.*

### *What do you think has been the most revolutionary discovery in human evolution studies over the last 5 years?*

Why, the fact that human evolution is so complex, of course! Hybridisation and chance have played a huge role in shaping hominin diversity (Ackermann, 2011; Ackermann et al. 2016, 2019; Schroeder et al. 2014; Schroeder and Ackermann 2017). But let's be honest, although this has received a lot of attention in recent years, the reality is that researchers – many from non-Western spaces – have been challenging simple models of hominin evolution (and especially human origins) for some time. We have simply gotten to a point where the genetics have supported previous hypotheses and made them more mainstream.

### *What is your favourite memory from your career?*

One of the happiest days of my life happened in September 2009, when I sat my then four PhD students (Riashna Sithaldeen, Lauren Schroeder, Tessa Campbell and Wendy Black... all South African women who have since completed) down in my office to tell them I was going to have a baby for the first (and only) time. I was 40, and quite anxious to be pregnant. I know that may not seem like an academic highlight, but the outpouring of sheer joy that came from them really drove home the fact that we had created this supportive and inclusive space together. They also assured me that having a boy was for the best as none of them would have wanted me as a mother (LOL). I have been deeply privileged to have the opportunity to know them, and the cohort of young South African palaeoscience students more broadly.

### *What are you most proud of?*

My students, right through from undergraduates to PhD students. I am proudest of them and everything they have achieved, some despite considerable adversity that people in the Western world can't fathom. I am especially proud of my PhD students, who are a beacon of hope in today's world.

*Rebecca with some of her current and former PhD students (left to right: Dr. Tessa Campbell, Dr. Lauren Schroeder, Dr. Riashna Sithaldeen, Rebecca, Robyn Humphreys (PhD candidate), Nomawethu Hlazo (PhD candidate), Dr. Kerryn Warren.*

### *What is your favourite thing about your job? What would you change if you could?*

I love giving students opportunities and watching them grow. I love the freedom and flexibility academia gives me, and them, to explore their ideas, and to change. I don't like the slow pace of social change in academia, and the fact that it is still largely a white man's world. I deeply dislike the continuation of practices that prop up the systemic inequalities that resulted from colonial practices (Ackermann, 2019). It bothers me immensely that helicopter research is still rampant, with Westerners bringing their money and people into African countries, in many cases with relatively little engagement with Africans as peers (and not just workers). I would change that in a minute.

***References:***


Ackermann, R. R. and Cheverud, J. M. (2004). Detecting genetic drift versus selection in human evolution. *Proceedings of the National Academy of Sciences of the United States of America* 101(52), 17946-17951. DOI: 10.1073/pnas.0405919102

Ackermann, R. R. and Cheverud, J. M. (2006). Identifying the morphological signatures of the hybridization in primate and human evolution. *Journal of Human* Evolution 51(6), 632-645. DOI: 10.1016.j.jhevol.2006.07.009

Ackermann, R. R. (2011). Phenotypic traits of primate hybrids: Recognizing admixture in the fossil record. *Evolutionary Anthropology: Issues, News and Reviews* 19(6), 258-270. DOI: 10.1002/evan.20288

Ackermann, R. R., Mackay, A. and Arnold, M. L. (2016). The hybrid origin of 'modern' humans. *Evolutionary Biology* 43(1), 1-11. DOI: 10.1007/s11692-015-9348-1

Ackermann, R. R. (2019). Reflections on the history and legacy of scientific racism in South African palaeoanthropology and beyond. *Journal of Human Evolution* 126, 106-111. DOI: 10.1016/j.jhevol.2018.11.007. Preprint AfricArxiv DOI: 10.31730/osf.io/t3v9a

Ackermann, R. R., Arnold, M. L., Baiz, M. D., Cahill, J., Cortez-Ortiz, L., Evans, B., Grant, B. R., Grant, P. R., Hallgrimsson, B., Humphreys, R., Jolly, C. J., Malukiewicz, J., Percivalm C. J., Ritzman, T., Roos, C., Roseman, C. C., Schroeder, L., Smith, F. H., Warren, K., Wayne, R. and Zinner, D. (2019). Hybridization in human evolution: insights from other organisms. *Evolutionary Anthropology* 28(4), 189-209. Preprint AfricArxiv DOI: 10.31730/osf.io/y3bp7

Athreya, S. and Ackermann, R.R. (2020). Colonialism and narratives of human origins in Asia and Africa. In: M, Porr and J, Matthews (eds), *Interrogating Human Origins: Decolonisation and the Deep Past.* Archaeological Orientation Series. Routledge: Abingdon. (Series editors: Christopher Witmore and Gavin Lucas). Preprint AfricaArxiv DOI: 10.31730/osf.io/jtkn2

Fuentes, A., Ackermann, R. R., Athreya, S., Bolnick, D., Lasisi,T., Lee, S-H., McLean, S. A. and Nelson. R. (2019). AAPA statement on race and racism. *American Journal of Physical Anthropology* 169(3), 400-402. DOI: 10.1002/ajpa.23882

Schroeder, L., Roseman, C. C., Cheverud, J. M. and Ackermann, R. R. (2014). Characterizing the evolutionary path(s) to early *Homo*. *PLoS ONE* 9(12): e114307. DOI: 10.1371/journal.pone.0114307

Schroeder, L. and Ackermann, R. R. (2017). Evolutionary processes shaping diversity across the *Homo* lineage. *Journal of Human Evolution* 111, 1-17. DOI: 10.1016/j.jhevol.2017.06.004. Preprint: BioRxiv DOI: 10.1001/13650

# Part 3: Earth science and palaeoclimatic change

## Professor Rick Potts[13]

Professor Rick Potts is the Director of the Human Origins Program at the Smithsonian National Museum of Natural History. Rick joined the Smithsonian in 1985 and has since focused his research toward understanding how Earth's environmental change affects early human adaptation. He formulated the well-received Variability Selection Hypothesis (Potts, 1996, 1998), proposing that hominin evolution responded to environmental instability, an idea that led him to develop many international collaborations among scientists interested in the ecological aspects of human evolution. Rick also leads excavations at early human sites in the East African Rift Valley, including the famous handaxe site of Olorgesailie, Kenya, and Kanam near Lake Victoria, Kenya.

*Photo credit: Smithsonian Human Origins Program.*

### *What are your research interests?*

I am a palaeoanthropologist with a PhD in Biological Anthropology. My main area is the long-term ecological history of human evolution, with a focus on behavioural adaptations to changing environments. Much of my work involves excavation at field sites in East Africa and China. So, the research I carry out depends on a stimulating fusion of evolutionary biology, palaeontology, archaeology, ecology, sedimentary geology, stratigraphy, geochronology, and diverse environmental sciences. I need to know as much as I can about these fields.

### *What originally drew you towards human evolution studies?*

The roots of my interest go back to my teenage years. For reasons I still don't understand, I was drawn from an early age to the origin of things: what were the predecessors of today's musical instruments, how did the rules of baseball develop, how did our solar system originate? Around 15 years of age, I began reading books about primate (including human) fossil discoveries. Once I found out I could explore the origin of 'us', I was hooked. It wasn't any particular television special or National Geographic article that captivated me. It was basic imagination – who were those early ancestors?

### *What was your PhD topic? How did you find your PhD experience?*

It still astonishes me I had the opportunity to know Dr. Mary Leakey, who (along with one of my thesis advisors, Dr. Alan Walker) paved the way for my thesis on early hominin activities and paleoecology at Olduvai. Another of my PhD supervisors, Dr. Erik Trinkaus, was (and is) incredibly fun to work with – and I assumed I would work on Neanderthal anatomy for my PhD. Yet Erik and Alan – and other graduate school mentors such as Steven Jay Gould – imparted a wondrous vision of the interdependent fields that has become present-day palaeoanthropology. The best way to begin was made feasible by Mary's permission (offered very warily) to study the fossil and tool remains from Bed I Olduvai. I was

[13] Smithsonian Institution Human Origins Program; pottsr@si.edu

lucky to have this type of PhD opportunity, an exhilarating time befitting my initial teenage imagination where it all started.

*Rick excavating an elephant tusk (around 1990). Photo Credit: Smithsonian Human Origins Program.*

### What has changed in academia since you did your PhD?

Let me answer the research side of this question first. The ease, speed, and breadth of communication has revolutionized the ability to convene worldwide groups of scientists, students, and the people who assist logistics and give permissions (leaders of local field crews, community leaders where we do fieldwork, international organizations, government officials, fund managers). Email and computational ability have made it feasible to transform individual or small teams into an opportunity to communicate with dozens of motivated collaborators and to find colleagues passionately devoted to solving questions on human evolution (for example, Potts et al., 2020). One has to be tenacious about cooperation and treat everyone fairly and with respect. The technology of communication, when used with goodwill, helps to get people together and motivated on shared scientific goals.

As for academia, like western society in general, there is a long way to go to assure significant leadership opportunities where diversity is lacking. There's much greater awareness about it since my graduate school years. The opportunities must also reach students and early career academics internationally, in countries where we work.

### *What current projects are you working on at the Smithsonian Institution? Where do you hope these go in the future?*

I appreciate the challenge of herding cats – that is, I delight in bringing teams of colleagues together on projects. For instance, more than 30 colleagues are working on a long sediment core drilled near Olorgesailie, in southern Kenya, very close to our excavations that uncovered a major shift in early human behaviour and ecological setting that began roughly 500,000 years ago. We're developing an incredibly precise ecological record of vegetation, water supply, and other things that mattered to how hominins and other mammals survived. The long-term goal is to inspire large teams of researchers to contribute to understanding the long-term ecological history of human evolution. I'd like to have a few years left to get that ball rolling down the hill – or, rather, making progress up that hill.

*Field team at Olorgesailie. Photo credit: Smithsonian Human Origins Program.*

### *What project or publication or achievement are you most proud of?*

Probably my 1996 book that no one has read – Humanity's Descent: The consequences of ecological instability (Potts, 1996). It was a 5-year project of researching, thinking, and writing. It was a delightfully lonely time that led to unexpected areas of thinking about evolutionary processes. The resulting concept of variability selection launched quite a few publications, head-shaking (the disapproving kind), and (I think) novel ideas about how ecological instability can lead to the evolution of adaptability. I consider adaptability to be an overarching theme in the study of human evolution. But I hadn't thought much about this theme until I took the time to learn and write about it.

*Rick and Muteti Nume excavating fossils and stone artifacts at Olorgesailie, Kenya. Photo credit: Smithsonian Human Origins Program.*

*Rick with Maasai students in South Kenya. Photo credit: Smithsonian Human Origins Program.*

***What is your favourite memory from the field?***

Without a doubt, sitting at the camp table at Olorgesailie with my friend Muteti Nume, the foreman of our Kenya field crew until 2018, when he passed away. We worked together for 35 years, spent every summer together. Over breakfast, we planned the day; at lunch, we barely spoke as the temperature climbed; by lantern light at dinner, we told stories of life, crops (his), memories, and hopes. Later on, it was great to have other researchers around the table; but those early days of fieldwork with Muteti are woven into the story of my life.

***If you were not an archaeologist/paleoanthropologist, what would you be?***

Possibly a school biology teacher exciting my students about photosynthesis and boring them about philosophy (how do we know things?). I'd have considered that a wonderful life.

***If you had a time machine, how far would you ask to go back, where would you go, and what would you want to see?***

I think if I went back (temporarily, of course) to one particular time, one particular place, to see one or several particular hominin species – I don't think I'd have a clue about what or who I was looking at! My guess is that the past is so different from my assumptions, any time (back far enough) would prove fascinating... and truly befuddling. In a time machine, I'd prefer to head at least 100 years into the future and be amazed by what the students and colleagues I'll never meet will have discovered and learned about our species' ancestry.

*Rick with students at the Human Origins Program. Photo credit: Smithsonian Human Origins Program.*

***References:***

Potts, R. (1996). *Humanity's Descent: The Consequences of Ecological Instability.* New York: William Morrow and Company.

Potts, R. (1998). Variability selection in hominid evolution. *Evolutionary Anthropology: Issues, News and Reviews* 7(3), 81-96. DOI: 10.1002/(SICI)1520-6505(1998)7:3<81::AID-EVAN3>3.0.CO;2-A

Potts, R., Dommain, R., Moerman, J. W., Behrensmeyer, A. K., Deino, A. L., Riedl, S., Beverly, E. J., Brown, E. T., Deocampo, D., Kinyanjui, R., Lupien, R., Owen, R. B., Rabideaux, N., Russell, J. M., Stockhecke, M., deMenocal, P., Tyler Faith, J., Garcin, Y., Noren, A., Scott, J. J., Wester, D., Bright, J., Clark, Cohen, A. S., Keller, C. B., King, J., Levin, N. E., Shanon, K. B., Muiruri, V., Renaunt, R. W., Rucina, S. and Uno, K. (2020). Increased ecological resource variability during a critical transition in hominin evolution. *Science Advances* 6, eabc8975. DOI: 10.1126/sciadv.abc8975

## Professor Mark Maslin[14]

Professor Mark Maslin is the Director of The London NERC Doctoral Training Partnership and Professor of Climatology at University College London. Mark is a leading scientist with particular expertise in the causes of past and future global climate change and its effects on the global carbon cycle, biodiversity, rainforests and human evolution. He has published over 170 papers in journals such as *Science, Nature, Journal of Human Evolution* and *The Lancet*, with a current citation count according to Google Scholar over 19,500 (H=65 and i10 index=162). He has written 10 books, over 60 popular articles and appears regularly on radio and television. His books include the high successful 'Climate Change: A Very Short Introduction' (OUP, 2014 and 2021), 'The Human Planet: How we created the Anthropocene' co-authored with Simon Lewis (Penguin, 2018), How to Save Our Planet: The Facts (Penguin 2021) and 'The Cradle of Humanity' (OUP, 2017 and 2019) which bring together the latest insights from hominin fossils, geology and palaeoclimatology to explore the evolution of our ultrasocial brains. He was included in Who's Who for the first time in 2009 and was granted a Royal Society Wolfson Research Merit Scholarship in 2011 for his work on human evolution.

### *What are your research interests and your particular area of expertise?*

My research interests are very wide, from human evolution to the development of the global green economy. I very much see myself as a natural scientist, using scientific methods to investigate important subjects such as human evolution, the Anthropocene, climate change and the other major challenges facing humanity in the 21st century. My areas of expertise can be summed up as understanding the fundamental causes of past and future climate change and their consequences for evolution, biodiversity, people and policy-making.

### *What originally drew you towards climatology?*

I have always been fascinated about how the world works and took Geology and Geography at University. However, I have a holistic view of the natural world and hence would describe myself as an Earth System scientist – because biology, climatology, ecology, biogeochemistry, oceanography, and geology are just some of the sciences we need to combine if we are to understanding how our planet works and our influence on it.

### *What was your PhD topic? How did you find your PhD experience?*

My PhD was at Cambridge University and supervised by the late Professor Sir Nick Shackleton FRS and Professor Ellen Thomas who is now at Yale University – both brilliant in their own ways. My PhD topic was on the palaeoceanography of the North Atlantic Ocean trying to understand quasi-cyclic collapses of the North American ice sheet during the last ice age. These so called 'Heinrich events' sent huge

[14] Department of Geography, University College London; m.maslin@ucl.ac.uk

armadas of icebergs crashing into North Atlantic Ocean disrupting the circulation of the deep ocean and affecting global climate.

The Cambridge PhD process was at that time very Darwinian – the survival of the fittest – there was a lack of regular supervision, no real official support, no one ever explained to me how one should approach a PhD or what were the expectations. But this experience was very valuable to me because when I become the Director of the London NERC Doctoral Training Partnership it meant I could develop a completely new way to training and supporting PhD students across the whole of London, simply by avoiding the failings of my own PhD training and by empowering students.

*A very excited Mark in the Omo National Park in south-western Ethiopia (2018) at the site where one of the earliest anatomically modern Homo sapiens was found and dated to 195,000 years ago. This is one of the most important sites in the study of human evolution.*

### *After your PhD, what positions have you held and where?*

After my PhD I was very lucky to get a couple of post-doctoral positions in marine geology and palaeoclimatology at Kiel University in Germany under the mentorship of Prof. Michael Sarnthein who trained a whole generation of brilliant scientists. It is also where my friendship and collaboration with Prof. Martin Trauth started and has led us to some startling findings regarding the causes of human evolution. After Kiel University, I was offered a position at UCL where I have stayed ever since. At UCL, I have had the privilege of being Head of the Geography Department, Director of the UCL Environment Institute and now Director of the London NERC DTP.

### *What are you currently working on? What do you hope to do in the future?*

This is probably the most difficult question to answer – as I have many different projects on the go in many different fields, from climate change health adaptation to the carbon footprint of coffee. One human evolution project I am very excited about is the work of one of my PhD student Cécile Porchier who is working on annually laminated diatom lake sediments from Kenya dated at 80 to 100 kyrs ago. She is co-supervised by colleagues at the Natural History Museum in London and if successful she will be able to understand past climate changes in East Africa at a yearly resolution to really understand what drove the evolution of modern humans and their dispersal out of Africa. I am also very excited as I am starting a new project called 'Human Evolution in the Anthropocene' with a friend and colleague Prof. Peter Kjærgaard, the Director of the Natural History Museum of Denmark and who knows where that project will take us.

*Travelling across the Chew Bahir palaeolake which lies between the Ethiopian and Omo-Turkana Rifts in 2018. It has been extremely dry for at least the last 100 years but was 20m deep around 5,500 years ago. First drilled in 2013/14, it has produced a stunning record of paleoclimate covering the last 650,000 years.*

### *What is your favourite memory from the field?*

My favourite memory from fieldwork was the first time we took the helicopter from our camp on the Rift shoulder and swooped down into the Suguta valley in Northern Kenya. It was then for the first time I really understood how the geology and the tectonics had created this amazing landscape and how changes in the Earth orbits could fill or drain these massive lake basins. There is also a strange feeling when one is camped on the Rift shoulders as the climate is perfect for humans – not too hot, not too cold with ample vegetation and water – it feels like home.

*Glorified fieldwork taxi (the helicopter) lands ready to take the next group of scientists into the Suguta Valley in. 2010. They starting research early in the morning before it gets too hot in the Valley, so travelling around by helicopter means that they can work from 7 am till ~3 pm and, if a site is a write-off, then they can move quickly to the next site of interest.*

### *What project or publication or achievement are you most proud of?*

In the field of human evolution, I think the work I am most proud of is the synthesis of all the data from East Africa and the realisation that the exciting story our evolution could only be understood by bringing lots of different research areas together. Martin Trauth and I realised that a combination of tectonics and orbital cycles created periods of time when deep freshwater lakes appeared and then disappeared within the East African Rift Valley (Maslin and Trauth, 2009). The climate cycled from extremely wet to very dry and coincided with major period of human evolution and dispersal. For example, the evolution of *Homo erectus*, *Homo heidelbergensis* and *Homo sapiens* and their dispersal out of Africa. We called this the 'Pulsed Climate Variability hypothesis' as it built on the work of Rick Potts (Potts, 1998) and provided a temporal framework within which human evolution could be understood. The central idea that peaks in precession forcing are linked through lakes to human evolution was radical 15 years ago when Martin and I suggested it but now it is so accepted that many forget to attribute it to the original research.

### *If you were not an earth scientist, what would you be?*

As an Earth System Scientist or a natural scientist, I do not really believe in the compartmentalism of science. I also do not believe in boundaries between science and social science and have worked on both. But as a second-year undergraduate student, I did a six-week internship in the summer with a leading international London Law firm – so I might have ended up being an environmental lawyer. Now that is a scary thought!

### *What is the best thing about your job and what is one thing you would change if you could?*

I have the best job in the world and people laugh when I say this – but I am serious. I am surrounded by some of the brightest people in the world both colleagues and students. I get to teach and train some of the most interesting students at all levels from undergraduate to PhD. I get to choose exactly what subjects I want to research, and no one worries when I stray far from my supposed areas of expertise to support students research the global green economy or food insecurity in Nicaragua or global health and climate change. My University is extremely supportive of public engagement and has allowed me to write 10 popular books (including 'The Cradle of Humanity') and many articles for New Scientist, Guardian, The Times and The Conversation. I also co-founded a company in 2012, Rezatec Ltd, which has grown to over 40 staff and a turnover of over £5 million per year. What other job would allow me to be teacher, trainer, mentor, researcher, author, presenter, explorer, and entrepreneur – and to work with the best of the best from every field of human endeavour.

### *References:*

Maslin M.A. and Trauth M.H. (2009). Plio-Pleistocene East African Pulsed Climate Variability and Its Influence on Early Human Evolution. In: Grine F.E., Fleagle J.G., Leakey R.E. (eds), *The First Humans – Origin and Early Evolution of the Genus Homo.* Vertebrate Paleobiology and Paleoanthropology. Springer, Dordrecht. DOI: 10.1007/978-1-4020-9980-9_13

Potts, R. (1998). Variability selection in hominid evolution. *Evolutionary Anthropology: Issues, News and* Reviews 7(3), 81-96. DOI: 10.1002/(SICI)1520-6505(1998)7:3<81::AID-EVAN3>3.0.CO;2-A

*Mark giving a Royal Society talk at the Cheltenham Science Festival in 2012 on the causes of early human evolution with hominin skull props!*

## Dr Yoshi Maezumi [15]

Dr Yoshi Maezumi is a palaeoecologist currently working at the University of Amsterdam! She is a Marie-Curie Fellow and National Geographic explorer, currently working on a project called: 'FIRE: Fire Intensity in Rainforest Ecotones'. Her research involves applying interdisciplinary approaches and methodologies to advance our understanding of long-term natural and anthropogenic paleoecological variability in the Neotropics. Yoshi also writes a blog called 'Her Science' which documents her experiences, adventures, and inspirations as a woman in science.

### *What are your research interests and your particular area of expertise?*

I study human impacts on past and present ecosystems using an interdisciplinary methodology combining archaeology, archaeobotany, palaeoecology, and palaeoclimatology. While my primary research centres on past crop cultivation, agroforestry, and fire management in the Amazon and Caribbean, I am also involved with diverse collaborative projects utilising similar multiproxy datasets to address questions of human-environment interactions in the United States, British Isles, Mediterranean, and Australia.

### *What originally drew you towards palaeoecology?*

I started University as a dance major. However, after a bad car accident that ended my professional dancing aspirations, I decided to go to college to be an Archaeologist. I completed a double BA in Anthropological Archaeology and Religious Studies. During this time, I conducted archaeological fieldwork in Jordan, Italy, Spain, and throughout the US. I completed a MA in Analytical Archaeology at CSU Long Beach. While conducting fieldwork in Guatemala, I had the opportunity to collect my first sediment cores in the mangroves near El Baul archaeological site and became increasingly interested in past human impacts on the environment. As I was learning palaeoecological proxies during my Masters, I went to train in charcoal analysis with Dr. Mitchell J. Power at the University of Utah who later became my PhD Supervisor.

### *What was your PhD topic and what were the findings from your PhD?*

My PhD topic was: Climate, Vegetation, and Fire Linkages in the Bolivian Amazon. During my PhD, I used multi-proxy analytical techniques to reconstruct long-term natural and anthropogenic drivers of palaeoecological change in the Bolivian Amazon. My research represented the first long-term palaeoecological study from the Amazon cerrado savanna ecosystem (Maezumi *et al.*, 2015).

[15] Institute for Biodiversity and Ecosystem Dynamics, University of Amsterdam; s.y.maezumi@uva.nl

*Examining rock art in Para, Brazil.*

### *After your PhD, where have you worked? Where has been your favourite place to work?*

Following my PhD, I held a 3-year Post-doctoral Research Fellowship at the University of Exeter, UK with Prof. Jose Iriarte. During my post-doc, I began to integrate my training in archaeology and palaeoecology to examine past human land use as a driver of ecological change. Implementing an interdisciplinary approach combining palaeoecology, archaeology, archaeobotany, palaeoclimatology and botany, our team published one of the first multidisciplinary, high-resolution reconstructions of past human land use and fire management in the Amazon.

Following my post-doc, I held a one-year Lectureship position at the University of the West Indies teaching courses in Palaeoclimatology and Environmental Change. During this time, I was awarded an Early Career National Geographic Grant for my project Jamaica a Last Island Frontier. This project investigates the impact of human colonization on island biodiversity and fire activity on the island. Jamaica is one of my favourite places in the world. The people are warm and friendly, the research potential is extraordinary, and the surf is excellent.

*Yoshi with a lake sediment core in Jamaica.*

***What current projects are you working on at the University of Amsterdam? Where do you hope these go in the future?***

Currently, my Marie Curie funded project FIRE: Fire Intensity in Rainforest Ecotones, investigates the role of ancient fire management in shaping the Bolivian rainforest-savanna boundary. Fire intensity (the maximum temperature of a fire) is a key component of post-fire recovery, however to date there is not a way to reconstruct fire intensity in past fire events. My Marie Curie research is aimed at developing a state-of-the-art method using Fourier-Transform InfraRed (FTIR) spectroscopy to chemically analyse fossil charcoal to provide the first proxy to reconstruct past fire intensity. This research will be used to evaluate long-term ecological impacts of past indigenous fire use. I am starting the job hunt for a permanent position. My 'dream-job' will enable me to continue my interdisciplinary research program and integrate this methodology into my teaching curriculum to train the next generation of interdisciplinary scientists.

### *What advice would you give to a student interested in your field of research?*

I think one of the things that opened-up the most doors for me was networking. I know not all students will have this luxury because of financial circumstances, but in each stage of my education, I was willing to pay out of pocket to travel to meet researchers I wanted to work with and attend workshops on things I wanted to learn that were not offered in my home department. This was how I met both my PhD and Post-doc supervisors and I met my current mentor, Dr. Will Gosling at a conference (the OSM/YSM meeting in Spain). We brainstormed the idea for my current Marie Curie project over coffee during that meeting. Additionally, I applied for every little pool of funding I could to help pay for my research and travel to conferences. Small grants for two hundred dollars here and five hundred dollars there really build up over time and looks great on your CV.

*Conducting fieldwork in the Amazon, collecting samples from the 2019 fire season.*

*Yoshi sampling sediment cores in Para, Brazil.*

***What do you think has been the most revolutionary discovery in human evolution studies over the last 5 years?***

Environmental DNA. To me, it's like a paleoecology time machine sprinkled with unicorn magic (I joke). But seriously, the advances in ancient DNA over the past few years has revolutionized our understanding of the palaeorecord.

### *If you were not a palaeoecologist, what would you be?*

I taught yoga for about 10 years (one of my jobs in grad school). I would love to continue teaching yoga if I was not a scientist. I think I almost have enough free time to start teaching yoga again, so maybe I'll get to do both!

### *References:*

Maezumi, S.Y., Power, M.J., Mayle, F.E., McLauchlan, K.K. and Iriarte, J. (2015). Effects of past climate variability on fire and vegetation in the cerrâdo savanna of the Huanchaca Mesetta, NE Bolivia. *Climate of the Past* 11, 835-853. DOI: 10.5194/cp-11-835-2015

# Part 4: Evolutionary anthropology and primatology

## Dr Duncan Stibbard-Hawkes[16]

Dr Duncan Stibbard-Hawkes is an evolutionary anthropologist. Duncan works with the Hadza in northern Tanzania, who have traditionally subsisted through hunting and gathering. He is interested in food-sharing, the use and abuse of signalling theory and forager egalitarianism. He previously won the Ruggles-Gates Award from the Royal Anthropological Institute as well as a grant from the Leakey Foundation for his PhD project: Reading the signals: What does Hadza hunting success honestly convey? Duncan recently finished a teaching fellowship at Durham University, where he lectured in evolutionary anthropology. He has just begun a postdoc jointly sponsored by the Universities of Durham and Pennsylvania for the 'Culture of Schooling' project, investigating Hadza engagement with formal education.

### *What are your research interests?*

My subject area is called 'hunter-gatherer studies'. Although there continues to be much debate about whether 'hunter-gatherer' even makes sense as a category, the name has stuck. I work with a population called the Hadza, who traditionally hunted and gathered for most of their food. I'm interested in the motivations underlying hunting and food-sharing. I'm interested in unknotting the reasons why, despite a spectrum of differences, there are some critical similarities between hunter-gatherer groups who are united by nothing other than a shared mode of subsistence. I'm interested in the ways by which forager populations adapt when traditional subsistence practices become less viable. Finally, I am interested in forager egalitarianism. Our history books are full of kings, queens, khans and emperors. How then, despite the manifold incentives to seize power, do populations like the Hadza manage so effectively to curtail attempts at aggrandisement and prevent people from naming themselves 'leaders'?

### *What originally drew you towards biological anthropology?*

Although we like to see ourselves as exceptional, humans are as much the products of evolution as any other species. And while there are many valid frameworks with which to view ourselves, no account of our actions, our minds and our forms is wholly complete without recourse to evolutionary logic. Given this fact, it is a constant source of consternation that evolutionary anthropology is not a larger or more well-known discipline. Why isn't evolutionary anthropology on the national curriculum? I enjoyed learning about the Tudors in school, but the origins of bipedalism are surely more universally elucidating than the English reformation. The Catcher in the Rye has a lot to say about the human condition, but perhaps not quite as much as The Origin of Species. So that's the draw of evolutionary anthropology. Humans do not make sense without it. Not completely.

How did I get into evolutionary anthropology in the first place though? The unglamorous truth is that it was an accident. I applied to Cambridge's now defunct 'Archaeology and Anthropology' course, with

[16] Department of Anthropology, University of Durham; duncanstibs@cantab.net

the idea of studying social anthropology. I was interested in learning about the full range of human experiences and cultures. As the course progressed, I did not always enjoy the sometimes high-minded epistemological wrangling of social anthropology, nor the unrelenting self-reflection and self-censure. Sometimes I felt I was learning more about what social anthropologists thought of each other than I was learning about the world and the people in it. At the same time, each lecture in the evolutionary anthropology course was a revelation. I remember reading Kristen Hawkes' original paper on the grandmother hypothesis one day in the library. It was such a clever and interesting piece of evolutionary logic that I decided then and there that I was sold. And that was that.

### *What was your PhD topic and what were the findings from your PhD?*

My PhD research examined the costly signalling hypothesis of human hunting, specifically the idea that hunters procure and share the meat from large animals as a way of showing-off their hunting skills. If hunting is a way of showing off, or signalling, then being known as a good hunter should be closely related to actually being a good hunter.

So, is it? I found that, among the Hadza, in aggregate, people's assessments of their peers' hunting skills were actually pretty accurate (Stibbard-Hawkes *et al.*, 2018). However, at an individual level, there was much error and noise. This is an example of a crowd wisdom effect. If you ask lots of people to guess the weight of a jar of sweets, the mean of their guesses is often freakishly close to the actual weight. But this can happen even when most individual estimates are pretty wide of the mark.

*Duncan's field vehicle parked under a thorny acacia tree.*

What do my results tell us about signalling? This is open to interpretation, but my personal take is that individual assessments of hunting ability are too error-prone for hunting and sharing to be a good way of signalling skill. But the good news is that aggregated reputation scores seem pretty accurate, so we should continue to use them as a proxy variable where actual skill is unknown!

My PhD research also addressed another question. There has been extensive debate about food-sharing and family provisioning. Do Hadza hunters share their food indiscriminately? Or do they keep the lion's share for their own families? I looked at the relationship between hunting reputation and nutrition and found that well-reputed hunters and their spouses had no better nutrition than did anyone else (Stibbard-Hawkes *et al.*, 2020). As discussed in the paper, it is difficult to prove a negative, and there are some finicky barriers to inference. But results are nonetheless consistent with generalised food-sharing, in line with previous reports by Nicholas Blurton Jones, Kristen Hawkes and James Woodburn.

### *Where did you complete your PhD and who was your supervisor?*

I did my PhD at Cambridge University. My supervisor was Frank Marlowe, who tragically took early retirement due to ill-health during my degree. His students wrote a collection of remembrances about his life and work for Human Nature which you can read here. When Frank retired, Robert Attenborough kindly took over as my supervisor, and continues to be a good friend and an occasional agony uncle.

### *What are you currently working on? What do you hope to do in the future?*

I currently have a few things in the works. The first piece of upcoming research reconsiders James Woodburn's theory that egalitarianism might be the consequence of democratised access to lethal weaponry – the idea that you shouldn't boss people around when they're armed and dangerous! I conclude that the evidence doesn't support this theory, but that it might hold some explanatory power in more limited contexts. In the second piece of research with Coren Apicella and Kris Smith, we asked many Hadza directly about what motivates them to hunt, to gather and to share food. Contrary to theoretical debates, most people, both men and women, highlighted that family provisioning and signalling were both important motivators for foraging work. Finally, I'm about to start work on a project looking at the changes brought about by increased participation in formal education.

Where do I hope this research will go in the future? I hope it will go into your endnote or Mendeley libraries!

### *What project or publication or achievement are you most proud of?*

My review article on Costly Signalling theory published in Evolutionary Anthropology (Stibbard-Hawkes, 2019) precipitated some friendly but occasionally forthright email exchanges with a couple of my academic heroes. However, I think the article raises some important questions. I would like to see greater opportunities for the interrogation of established theories and frameworks by young scholars and I was very grateful to the editor for giving me the chance to publish these ideas. However, the article I am proudest of is 'A Noisy Signal' (Stibbard-Hawkes *et al.*, 2018), which I have discussed above.

### *What do you think has been the most revolutionary discovery in your field over the last 5 years?*

In my own field of human behaviour, with a few exceptions, I think things most often progress through a process of gradual evolution and not revolution. It's pretty difficult to dig up a new behaviour! My top four papers from the last five years are Sceleza *et al.* (2020)'s eye-opening recent paper on diversity in human reproductive strategies, Ringen *et al.* (2019) and Ember et al. (2018)'s cross-cultural investigations of the association between risk/resource stress and food-sharing, and Singh *et al.* (2017)'s review paper on self-interested norm enforcement. Closer to home, Alyssa Crittenden's group have published some really good recent research about recent Hadza dietary changes, e.g., Pollom *et al.* (2020)'s just-published paper about how a mixed-subsistence diet might actually have some advantages over a purely foraged one. I mentioned the importance of interrogating established theories, and I also want to highlight how much I liked Dan Smith's recent paper on cultural group selection. It's super compelling. Check it out.

*Taking shelter from the midday sun in the shade of a baobab.*

### *What advice would you give to a student interested in your field of research?*

The easy answer is that you should follow your passions. The harder answer is that a research career can be stressful. PhDs can, for some students, feel like doing low-paid work for your supervisor. Moreover scholarship, especially anthropology, is an oversaturated industry and the number of qualified applicants exceeds the number of jobs. So, make sure you know the downsides, make sure you've talked to people and make sure you're going in with your eyes open. Make sure you have a backup plan. And if you've done all of that then go ahead and follow your passions!

## *What would you be if you were not an evolutionary anthropologist?*

One of my greatest regrets in pursuing anthropology was that I had to abandon my nascent professional wrestling career. Though if not a wrestler, probably a journalist, maybe a foreign correspondent.

*Duncan and the 2015 field team, Charles and Ibrahim.*

**References:**


Ember, C. R., Stoggard, I., Ringen, E. J. and Farrer, M. (2018). Our better nature: Does resource stress predict beyond-household sharing? *Evolution and Human Behaviour* 39(4), 380-391. DOI: 10.1016/j.evolhumbehav.2018.03.001

Pollom, T. R., Cross, C. L., Herlosky, K. N., Ford, E. and Crittenden, A. N. (2020). Effects of a mixed-subsitence diet on the growth of Hadza children. *American Journal of Human Biology* 33(1). DOI: 10.1002/ajhb.234555

Rigen, E .J., Duda, P. and Jaeggi, A. V. (2019). The evolution of daily food sharing: Bayesian phylogenetic analysis. *Evolution and Human Behaviour* 40(4), 375-384. DOI: 10.1016/j.evolhumbehav.2019.04.003

Scelza, B. A., Prall, S. P., Swinford, N., Gopalan, S., Atkinson, E. G., McElreath, R., Sheehama, J. and Henn, B. M. (2020). High rate of extrapair paternity in a human population demonstrates diversity in human reproductive strategies. *Science Advances* 6(8), eaay6195. DOI: 10.1126/sciadv.aay6195

Singh, M., Wrangham, R. and Glowacki, L. (2017). Self-Interest and the Design of Rules. *Human Nature* 28, 457-480. DOI: 10.1007/s12110-017-9298-7

Stibbard-Hawkes, D. N. E. (2019). Costly signalling and the handicap principle in hunter-gatherer research: a critical review. *Evolutionary Anthropology: Issues, News and Reviews* 28(3). DOI: 10.1002/evan.21767

Stibbard-Hawkes, D. N. E., Attenborough, R. D., Marlowe, F. W. (2018). A noisy signal: To what extent are Hadza hunting reputations predictive of actual hunting skills? *Evolution and Human Behaviour* 19(6), 639-651. DOI: 10.1016/j.evolhumbehav.2018.06.005

Stibbard-Hawkes, D. N. E., Attenborough, R. D., Mabulla, I. A., Marloew, F. W. (2020). To the hunter go the spoils? No evidence of nutritional benefit to being or marrying a well-reputed Hadza hunter. *American Journal of Physical Anthropology* 173(1). DOI: 10.1002/ajpa.24027

## Dr Ammie Kalan[17]

Dr Ammie Kalan is a primatologist at the Max Planck Institute for Evolutionary Anthropology. Ammie is a postdoctoral researcher investigating chimpanzee culture and communication as part of the Pan African Project: The Cultured Chimpanzee. Over her career as a primatologist, she has conducted fieldwork in Guinea-Bissau, Tanzania, Côte d'Ivoire, Republic of Congo and Costa Rica. She has developed a passive acoustic monitoring system for primates living in tropical forests and continues to be interested in bridging the gap between behavioural research and applied conservation through the use of non-invasive monitoring. Next year, she will be starting as a tenure-track Assistant Professor in Anthropology at the University of Victoria in British Columbia, Canada.

***What are your research interests and your particular area of expertise?***

I am a primatologist who specializes in great ape behavioural ecology, with a particular interest in tool use, culture and communication. I also actively work on improving remote methods used to study wild primates, not just great apes, in the field, namely passive acoustic monitoring and camera-trapping.

*Ammie biking to the remote camp site in Boé after a brief visit to a nearby village to stock up on supplies and use electricity. Photo credit: Ammie Kalan.*

[17] Max Planck Institute for Evolutionary Anthropology and Department of Anthropology, University of Victoria; ammie_kalan@eva.mpg.de

### *What originally drew you towards primatology?*

When I was in grade 5, so about 10 years old, I remember learning about endangered species, particularly the mountain gorilla and having the feeling that I wanted to do something to help and to know these creatures better. Not too long after I remember watching in awe as David Attenborough got up close and personal with wild mountain gorillas on an episode of BBC Life (I think?) and never being able to forget that remarkable moment. Growing up in the beautiful Pacific Northwest I knew I wanted to dedicate my career to wildlife ecology and/or environmental conservation but I could not get these wild great apes out of my head. So after studying Zoology for my BSc I volunteered for my first experience studying wild primates in Costa Rica as a research assistant and soon after applied for a Masters program that specialized in Primate Conservation at Oxford Brookes University in the UK. After moving to Oxford for this masters I was able to have my very own run in with wild gorillas (western lowland gorillas) when I conducted my first field research in Africa at the Lac Tele Reserve in the Republic of Congo for my dissertation project (Kalan and Rainey 2009; Kalan et al. 2010).

*Ammie recording food calls and pant hoot vocalisations of the Taï chimpanzees in the Côte d'Ivoire. Photo credit: Ammie Kalan.*

### *What was your PhD topic? How did you find your PhD experience?*

My PhD was on the social and ecological context of chimpanzee acoustic communication and its potential for biomonitoring. It was essentially two projects simultaneously since my supervisors were concerned by the riskiness of my biomonitoring project that they wanted to make sure I would have a backup plan, so to speak. This suited me just fine because I got the best of both worlds, I was able to do a project that was novel, technically challenging, but would be pioneering if it worked and also be able to follow habituated chimpanzees on foot to record their vocalizations and try to better understand what they might be communicating to one another. In the end, the biomonitoring project was generally a success, since I was able to show how remote audio recording units can be installed in a forest to record wild primate sounds that can later be extracted from these continuous forest recordings using semi-automated algorithms (Heinicke *et al.*, 2015) and thereby provide biologists with large-scale data on primate presence in an area (Kalan *et al.*, 2015b, 2016). I was also able to make some interesting new observations about chimpanzee food calls (Kalan et al., 2015a; Kalan and Boesch, 2015) and pant hoot vocalizations (Kalan and Boesch, 2018) with the data I recorded following individual chimpanzees, not to mention the incredible privilege to be able to get to know individual chimpanzees as they tolerated my presence from the moment they woke up until they built their evening nests.

***What current projects are you working on? Where do you hope these go in the future?***

I am currently a postdoctoral research for the Pan African Programme based at the Max Planck Institute for Evolutionary Anthropology in Leipzig, Germany. Here I work on integrating behavioural data we have collected from over 40 PanAf temporary research sites to better understand the ecological and environmental drivers and threats to chimpanzee behavioural and cultural diversity (Kühl et al. 2019; Kalan et al. 2020). Much of this work has already helped me to establish new collaborations and to start my own research projects, such as chimpanzee accumulative stone throwing based in Boé, Guinea-Bissau (Kalan *et al.*, 2019) with the support of an NGO, the Chimbo Foundation. In the future I will continue to use the PanAf dataset as a means to investigate questions that have thus far been difficult to answer using only a handful of populations.

*Overview of the PanAf project's methodological approach. Photo credit doc.station Medienproduktion GmBh, http://panafrican.eva.mpg.de/english/press.php.*

### *Why is your research important for understanding human evolution?*

Studying great apes such as chimpanzees provides us with the unique opportunity to observe and investigate the characteristics of a species closely related to humans that is still living today. The same cannot be said for the many fossils we have of our hominin ancestors, therefore primatology in general can provide great comparative insight into the course of human evolution given that we too are primates. For example, if we consider that the chimpanzee or *Pan* lineage split from the *Homo* lineage approximately 7-8 million years ago, then by comparing these two lineages today we can hypothesize which traits would have been present in our last common ancestor.

*Ammie following habituated chimpanzees of the Taï Chimpanzee Project. They wear face-masks and maintain a distance of 10m to protect the chimpanzees from potential diseases that they may carry and vice versa. Photo credit: Ammie Kalan.*

### *What is your favourite memory from the field?*

It's very difficult to pick one but a particular moment does stick out. While I was following the habituated chimpanzees of the Taï Chimpanzee Project there was one day when the individual I was following met up with other members of the group at a nice, open nut cracking site. As we arrived, I sat on a dead log to rest my feet while still observing my focal and as I did so one of the adolescent males at the time, called Ibrahim, started to walk directly towards me. Now Ibrahim was a bit naughty, as young male chimpanzees going through puberty often are, where he often displayed curiosity and interest in the human researchers and would at times come too close to us. So, I was a bit apprehensive about what his intentions were this time since I was well aware of the fact that I was in a sitting position and we were essentially eye to eye. As he got closer, I noticed he had a small stick in his mouth and was playfully running up to me. He then stopped about just a meter short of me and threw the small stick at my feet while making a lovely play face: he had just invited me to start playing with him using an object invitation as chimps commonly do amongst each other. I felt very honoured and happy inside but could not display any response since this is how we maintain a healthy separation from the chimps. He waited for me to do something and did look a bit disappointed when I didn't move an inch but then quickly turned his attention to another more playful participant that had followed him. I could then move away from them to maintain a better distance. It took all my restraint to not react to his invitation that day.

***What is the best thing about your job and what is one thing you would change if you could?***

I love that I get to work in the field, immersed in wild places and wild animals, where essentially anything can happen. It can be a little daunting at times, even scary or terrifying in some of the places/ circumstances in which I've worked, but it is never boring or repetitive. You never know what discovery awaits you when you are in the field, and the excitement that comes from experiencing that is well worth the strenuous physical and psychological toll that being in remote, isolated places can bring. The one thing I would change if I could is that there would be fewer tiny insects to contend with while collecting data, or at least that they would ignore humans as potential prey. For example, the little flies that go right into your eyes so that you literally cannot see, or the flies that lay eggs under your skin, or the ants that in a second will get underneath all your clothes and painfully bite you all over....

*Kuba, an adult male chimpanzee of the South group Taï chimpanzees. Photo credit: Ammie Kalan/Taï Chimpanzee Project.*

***If you were not a primatologist, what would you be?***

I would've become an environmental scientist, a route that I once seriously considered before choosing to do my masters degree in primate conservation. I was particularly fond of tropical forest ecology and interested in assessing and mitigating the effects of human disturbance on natural ecosystems so I'm sure I would have ended up not too far from what I am doing now.

***References:***


Heinicke, S., Kalan, A. K., Wagner, O. J. J., Mundry, R., Lukashevich, H. and Kühl, H. S. (2015). Assessing the performance of a semi-automated acoustic monitoring system for primates. *Methods in Ecology and Evolution* 6:753–763. DOI: 10.1111/2041-210X.12384

Kalan, A. K. and Boesch, C. (2015). Audience effects in chimpanzee food calls and their potential for recruiting others. *Behavioral Ecology and Sociobiology* 69:1701–1712. DOI: 10.1007/s00265-015-1982-1

Kalan, A. K. and Boesch, C. (2018). Re-emergence of the leaf clip gesture during an alpha takeover affects variation in male chimpanzee loud calls. PeerJ 6:e5079. DOI: 10.7717/peerj.5079

Kalan, A. K., Carmignani, E., Kronland-Martinet, R., Ystad, S., Chatron, J. and Aramaki, M. (2019) Chimpanzees use tree species with a resonant timbre for accumulative stone throwing. *Biology Letters* 15:20190747. DOI: 10.1098/rsbl.2019.0747

Kalan, A. K., Kulik, L., Arandjelovic, M., Boesch, C., Haas, F., Dieguez, P., Barratt, C. D., Abwe, E. E., Agbor, A., Angedakin, A., Aubert, F., Ayimisin, E. A., Bailey, E., Bessone, M., Brazzola, G., Buh, V. E., Chancellor, R., Cohen, H., Coupand, C., Curran, B., Danquah, E., Deschner, T., Dowd, D., Eno-Nku, M., Fay, J. M., Goedmakers, A., Granjon, A. C., Head, J., Hedwig, D., Hermans, V., Jeffery, K. J., Jones, S., Junker, J., Kadam, P., Kambo, M., Kienast, I., Kujirakwinja, D., Langergraber, K. E., Lapuente, J., Larson, B., Lee, K. C., Leinert, V., Llana, M., Marrocoli, S., Meier, A. C., Morgan, B., Morgan, D., Neil, E., Nicholl, S., Normand, E., Ormsby, L. J., Pachesco, L., Piel, A., Preece, J., Robbins, M. M., Rundus, A., Sanz, C., Sommer, V., Stewart, F., Tagg, N., Tennie, C., Vergnes, V., Welsh, A., Wessling, E. G., Willie, J., Wittig, R. M., Yuh, Y. G., Zuberbühler, K. and Kühl, H. S. (2020) Environmental variability supports chimpanzee behavioural diversity. *Nature Communications* 11:4451. DOI: 10.1038/s41467-020-18176-3

Kalan, A.K., Madzoké, B. and Rainey, H.J. (2010). A preliminary report on feeding and nesting behavior of swamp gorillas in the Lac Télé Community Reserve. *Mammalia* 74(4), DOI: 10.1515/mamm.2010.047

Kalan, A. K., Mundry, R. and Boesch, C. (2015a) Wild chimpanzees modify food call structure with respect to tree size for a particular fruit species. *Animal Behaviour* 101:1–9. DOI: 10.1016/j.anbehav.2014.12.011

Kalan, A. K., Mundry, R., Wagner, O. J. J., Heinicke, S., Boesch, C., Kühl, H. S. (2015b) Towards the automated detection and occupancy estimation of primates using passive acoustic monitoring. *Ecological Indicators* 54:217–226. DOI: 10.1016/j.ecolind.2015.02.023

Kalan, A. K., Piel, A. K., Mundry, R., Wittig, R. M., Boesch, C. and Kühl, H. S. (2016) Passive acoustic monitoring reveals group ranging and territory use: a case study of wild chimpanzees (Pan troglodytes). *Frontiers in Zoology* 13:34. DOI: 10.1186/s12983-016-0167-8

Kalan, A.K. and Rainey, H.J. (2009). Hand-clapping as a communicative gesture by wild female swamp gorillas. *Primates* 50, 273–275. DOI: 10.1007/s10329-009-0130-9

Kühl, H. S., Boesch, C., Kulik, L., Haas, F., Arandejelovic, M., Dieguez, P., Bocksberger, G., McElreath, M. B., Agbor, A., Angedakin, S., Ayimisin, E. A., Bailey, E., Barubiyo, D., Bessone, M., Brazzola, G., Chancellor, R., Cohen, H., Coupland, C., Danquah, E., Deschner, T., Dowd, D., Dunn, A., Egbe, V. E., Eshuis, H., Goedmakers, A., Granjon, A-C., Head, J., Hedwig, D., Hermans, V., Imong, I., Jeffery, K. J., Jones, S., Junker, J., Kadam, P., Kambere, M., Kambi, M., Kienast, I., Kujjrakwinja, E., Langergraber, K. E., Lapuente, J., Larson, B., Lee, K., Leinert, V., Llana, M., Maretti, G., Marrocoli, S., Martin, R., Mbi, T. J., Meier, A. C., Morgan, B., Morgan, B., Morgan, D., Mulindahabi, F., Murai, M., Neil, E., Niyigaba, P., Ormsby, L. J., Orume, R., Pacheco, L., Piel, A., Preece, J., Regnaut, S., Rundus, A., Sanz, C., van Schijndel, J., Sommers, V., Stewart, F., Tagg, N., Vendras, E., Vergnes, V., Welsh, A., Wessling, E. G., Willie, J., Wittig, R. M., Yuh, Y. G., Yurkiw, K., Zuberbühler, K. and Kalan, A, K. (2019) Human impact erodes chimpanzee behavioral diversity. *Science* 363:1453–1455. DOI: 10.1126/science.aau4532

## Professor Lynne Isbell[18]

Professor Lynne Isbell is a primatologist at the University of California, Davis (UC Davis). Lynne currently holds the positions of Professor and Chair of the Department of Anthropology at UC Davis and is President of the American Society of Primatologists. Her research program is focused on primate socioecology, particularly on aspects of food competition, predation, dispersal, and ranging behaviour. She is field-oriented and has engaged in multi-year fieldwork in Uganda and Kenya, with briefer forays into Madagascar, Tanzania, Rwanda, and the Democratic Republic of Congo. She also is the author of the award-winning book, The Fruit, the Tree, and the Serpent: Why We See So Well (Isbell, 2009).

***What are your research interests and your particular area of expertise?***

I am broadly interested in animal behaviour and my specialty within that is primate socioecology, the study of how ecology influences primate social organizations. In addition to observing the animals themselves in their natural habitats, I have explored the qualities of food that influence intragroup and intergroup competition, and relationships between primates and their predators.

***What originally drew you towards primatology?***

Jane Goodall started me on that path. I had always been interested in animals and animal behaviour but when I saw the cover of her book, In the Shadow of Man, that showed a young woman much like me following chimpanzees in a forest, I immediately bought it and read it. Before that, I thought the only way to work with animals was as a veterinarian. As an undergraduate, my opportunities were with ungulates such as captive bongos and desert bighorn sheep. My focus on primates began after I moved to Davis and volunteered to work on a year-long behavioural project with captive bonnet macaques. I was hooked! Primates are so much more active than ungulates!

***What was your PhD topic? How did you find your PhD experience?***

My dissertation title was 'Influences of Predation and Resource Competition on the Social System of Vervet Monkeys (*Cercopithecus aethiops*) in Amboseli National Park, Kenya.' It wasn't what I set out to do but I took advantage of a unique opportunity. It's very difficult to plan a study of the effects of predation on prey animals but when leopards kept eating my study animals toward the end of a slow die-off of the trees that vervets prefer to feed from and sleep in, it gave me a wonderful chance to explore the relative importance of predation and food resources on the lives of vervets.

My Ph.D. experience was one of the best times in my life. I was older when I started grad school and by then I knew it was the right place for me. But I also had the perspective that if I stopped enjoying it, I could always leave and do something else with my life. That mindset gave me a sort of freedom from grad school's often oppressive structure. I'd spent two years in the field in Uganda before I started grad school and my intention at first was to write my dissertation on aspects of red colobus monkey behaviour and ecology, but I had been bitten by 'the Africa bug' and when the opportunity arose to go

[18] Department of Anthropology, University California Davis; laisbell@ucdavis.edu

back for another couple years to follow vervets around, I jumped at it. I will always be grateful to so many people for giving me a chance: Tom Struhsaker, who taught me field methods; Peter Rodman, who took me on as his grad student; Meredith Small, who, one day while offering me my first ever academic job as a Reader for her class, looked me in the eye and said, 'you're good', something no one had ever said to me, and; Robert Seyfarth and Dorothy Cheney, who trusted me to be their field manager. I'm not even mentioning those who made my Ph.D. experience possible in other ways but they know who they are, and I thank them, too.

*Lynne standing in front of the tent where she stayed for a time in Amboseli National Park, Kenya, when she conducted her dissertation research (1986).*

### *After your PhD, what positions have you held and where?*

Right after I earned my degree, I was hired as the lab coordinator for a huge introductory organismal biology course at UC Davis. I was in charge of the labs and of 35 TAs who taught the lab sections, and I had to make sure they knew the material. I did that for four academic quarters. Then I was offered a tenure-track position at Rutgers University in New Jersey, which I held for three years before being invited back to UC Davis.

## *What current projects are you working on? Where do you hope these go in the future?*

I have to say that my research has ended up taking a back seat to my administrative work as Chair of my department these past several years. So, I'm eager to wrap up my last year as Chair and get back into the literature. Does that count as a project? What I'd like is to have the time to think deeply about what I'm reading and make connections that I didn't know were there before. Then I'd like to write a book or two about those connections, preferably in beautiful places such as on the Laikipia Plateau in Kenya and in the Rocky Mountains in Colorado. I think it would be a good use of my sabbatical time.

## *What project or publication or discovery are you most proud of?*

I really can't narrow it down to just one. Three are equally weighted, all taking many years of my time and energy but in very different ways. After working at two field sites that others had developed, I was really proud to be able to develop and maintain my own long-term field site at Segera in Kenya where my students and I could study the behaviour and ecology of vervets and patas monkeys, the latter being very difficult animals to habituate to human presence. That field project ran for 10 years.

*Lynne following habituated patas monkeys during a long-term field project at Segera Ranch, in Kenya (1999).*

I am also proud of developing the Snake Detection Theory (Isbell, 2006). Once the question popped into my head — could snakes have favoured the origin of primates and the subsequent diversity of the major primate clades? — it took about 10 years to investigate and understand the literature from multiple disciplines about which I knew next to nothing at first, in order to convince myself that there really was something to that question, and then to think about what I was reading in order to synthesize it into a coherent theory, and finally to write it all down so that others could see what I was seeing.

*Lynne with a female leopard that they had finally trapped and anesthetized to put on a GPS collar at Mpala Research Centre, Kenya (2014). The goop in the leopard's eyes prevented them from drying out.*

Finally, I'm proud of one particular field study that on the surface lasted one year, but, in fact, took 30 years to complete. As I already mentioned, leopards decimated my study groups in Amboseli when I was conducting my dissertation research. I'd wanted at that time to put radio-collars on the leopards I kept seeing very fleetingly, but it didn't happen. Then, 14 years later at my Segera field site leopards did the same thing again, and yet we were no wiser about leopard/primate interactions because leopards wait until we're gone to do their killing. Another 13 years passed and then GPS technology and grant support made it possible to remotely investigate in fine detail how GPS-collared leopards interact with GPS-collared olive baboons and vervets. Logistically it was a very difficult project to initiate but with a great team in the field, including Laura Bidner, Dairen Simpson, Mathew Mutinda, George Omondi, and Wilson Longor, we pulled it off with surprising results, e.g., leopards spent a lot more time near vervets than baboons but baboons were at greater risk when leopards were nearby, and when leopards hunted, they killed vervets during the day, but baboons, at night (Isbell *et al.*, 2018).

### *What is your favourite memory from the field?*

My first thoughts turned to events that are stuck in my mind because they were scary. They're good stories to tell but I wouldn't call them favourite memories. For a happy favourite memory, I always smile in awe when I think back to the moment when I stood in the dry river bed of the Mutara River at Segera and turned to see the beginning of the fresh new river trickle toward me from its origin as rain in the Aberdare Mountains. How many people ever get to see the front of a river? I am a very lucky person! The times I touched a leopard were also very special, almost spiritual, moments for me.

### *If you were not a primatologist, what would you be?*

I think maybe an investigative reporter. I like the truth and I enjoy searching for clues to puzzles.

### *If you had a time machine, how far would you ask to go back, where would you go, and what would you want to see?*

It would be really cool to see *Australopithecus* (*Paranthropus*) *boisei* in the wild and to study its behaviour and ecology.

### *References:*


Isbell, L. A. (2006). Snakes as agents of evolutionary change in primate brains. *Journal of Human Evolution* 51(1), 1-35. DOI: 10.1016/j.jhevol.2005.12.012.

Isbell, L. A. (2009). *The Fruit, the Tree and the Serpent: Why We See So Well.* Harvard University Press.

Isbell, L.A., Bidner, L.R., Van Cleave, E.K., Matsumoto-Oda, A., and Crofoot, M.C. (2018). GPS-identified vulnerabilities of savannah-woodland primates to leopard predation and their implications for early hominins. *Journal of Human Evolution* 118, 1-13. DOI: 10.1016/j.jhevol.201802.003

# Part 5: Human disease co-evolution

## Dr Hila May[19]

Dr Hila May is a physical anthropologist based at the Department of Anatomy, Sackler Faculty of Medicine and the Dan David Center, Tel Aviv University. Hila leads the Biohistory and Evolutionary Medicine Laboratory at Tel-Aviv University, which has two principal fields of interest: 1) the evolutionary trade-offs between different anatomical structures during an evolutionary process of adaptation, and their impacts on modern human health and 2) the reconstruction of the everyday lives of past population through their skeletal remains. She has appeared many times in the media discussing the significance of new discoveries, such as the jawbone from Misiliya Cave. She also has published in a number of high-impact academic journals, such as *Nature*, *Science* and *Journal of Human Evolution.*

### *What are your research interests and your particular area of expertise?*

I like to think about my research as multidisciplinary, involving human evolution, biohistory, and evolutionary medicine topics. Each of them stands on its own but they are related and complementary. To summarize it in a nutshell, I would say that my research focusses on five major issues:

1. Current human health in light of biological, cultural and technological evolution throughout the Pleistocene and Holocene.
2. The reconstruction of daily life in prehistoric populations, including issues relating to demography, physical activity, diet, health, group violence (intra and inter), labour division, and social behaviour.
3. The effect of technological revolutions on human biological structure, mainly the Agriculture revolution (ca. 15k years ago) and the Secondary products revolution (ca. 8k years ago).
4. The origin of Levantine prehistoric and historic populations based on ancient DNA.
5. Improving methodologies and creating new research tools for studying skeletal remains (e.g., methods for sexing skeletal remains and diagnosing pathologies).

### *Why did you originally want to study physical anthropology? And tell us about your PhD?*

Actually, before starting my MA degree I wasn't even aware of physical anthropology, and like all good stories, I found it by chance. You see, when I started my studies at the university, I could not decide between science or humanities, so I postponed my decision by combining the two and graduated as a BSc in Life sciences and Sociology and Anthropology. After graduating my BSc degree, my lack of decisiveness still had a strong grip, so I started to look for researchers that combine biology and anthropology and who use biological methods. This search mission turned out to be not an easy one, as there are only a few scholars in Israel that are carrying out this type of research.

[19] Department of Anatomy, Sackler Faculty of Medicine and the Dan David Center, Tel Aviv University; mayhila@tauex.tau.ac.il

By my mere luck, I came across Prof. Israel Hershkovitz from the Sackler Faculty of Medicine, Tel Aviv University who became my supervisor in my master and PhD theses. It was him that gave a name to what I was passionate about, and exposed me to the wonderful, exciting, and never dull worlds of human evolution, physical anthropology, and evolutionary medicine. My MSc was in evolutionary medicine where I studied an interesting phenomenon named HFI – Hyperostosis Frontalis Interna (an overgrowth in the inner part of the frontal bone) that's etiology is most probably related to sex hormones and has increased significantly during the last century (as I demonstrated in my study). My PhD thesis was in physical anthropology where I focused on the impact of the Agriculture Revolution in the Levant (ca. 15k years ago) on human biology from various aspects including nutrition, physical load and health.

*Hila excavating a burial in Timna as part of the Central Timna Valley project of Tel Aviv University, located in the southern Aravah, Israel.*

## *After your PhD, what positions have you held and where?*

Before starting my position as principal investigator at Sackler Faculty of Medicine, Tel Aviv University, I was a post doctorate fellow at the Institute of Evolutionary Medicine at the University of Zurich, Switzerland under the supervision of Prof. Frank Ruhli. During that period, I expanded my knowledge and skills in virtual anthropology methods and carried out some exciting projects in evolutionary medicine. For this project, I developed a new protocol for capturing the shape of the femur for revealing how changes in femoral shape during human evolution is related to the risk for a hip fracture.

### *What are you currently working on? What do you hope to do in the future?*

In the last years, I am involved in several projects studying new fossils exposed in Israel (Hershkovitz *et al.*, 2014; Hershkovitz *et al.*, 2018; Borgel *et al.*, 2019) as well as in a new excavation where fossils dated to ca. 100kya were exposed. These fossils have an enormous importance for understanding modern human evolution and dispersal out of Africa. I also lead several studies that aim to reveal the mysteries of past populations, such as their origin (through aDNA studies) (Harney *et al.*, 2018), physical activity (May and Ruff, 2016), health (Floreanova *et al.*, 2019), diet (May *et al.*, 2018; Pokhojaev *et al.*, 2019), demography, and social behaviour. These studies have great impact on the understanding of human adaptation to new environments throughout time. My focus is on populations who lived during the Agriculture Revolution and the Secondary Product Revolution in the Levant. Both these transitional periods had a major effect on human daily life and resulted in a modified environment (e.g., physical activity, diet, and population density) which eventually developed into the modern one, as we know it today. Other projects that are currently being carried out in my lab try to reveal, from an evolutionary perspective, why modern humans suffer from certain diseases such as osteoarthritis, osteoporosis, and hip fractures. Understanding the evolutionary causes of these diseases may help in finding ways for preventing them. Furthermore, in my lab, a large effort is placed on developing new methods that will enable us to answer our research questions as well as to enlarge the repertoire of methods for osteological studies (Sella Tunis *et al.*, 2017, 2018; Peleg *et al.*, 2020).

The great thing of being an anthropologist in Israel is that exciting findings never end. I plan to keep being involved in many other archaeological sites, expose new osteological findings, either from the far past or more recent, and to increase the number of the puzzle pieces at hand, for improving our understanding of human biology and behaviour.

### *What do your roles at the Department of Anatomy, Sackler Faculty of Medicine and the Dan David Centre entail on an average day?*

During an average day, I wear and replace between several hats. I have a great research group composed of about 10 MSc and PhD students that I supervise. I teach anatomy to medical students as well as courses in biological anthropology and human evolution for graduate students. In addition, I am a curator in the Anthropological collection in Dan David Center for Human Evolution and Biohistory Research, Tel Aviv University where I am responsible for the preservation laboratory and, of course, the collection itself. Finally, my professional activity also includes being a leading anthropologist in various archaeological sites in Israel (sites that vary in their age from 100K to 2K years) so from time to time I spend my days in the field.

*Entering Safsuf Cave; a chalcolithic cave at the Upper Galilee, Israel.*

*Peqi'in Cave; a Chalcolithic secondary burial cave at the Upper Galilee, Israel.*

### *Why is your research important for understanding prehistoric human behaviour?*

Skeletal remains contain objective information regarding the individual biology and behaviour, which enable us to reconstruct past populations osteobiographies. Moreover, each peace we add to the puzzle of human evolution and history improve our understanding of how we got here and why our body is designed the way it is. By applying various methods, both traditional and novel (the latter are mostly based on imaging techniques like microscopy, micro-CT, and virtual anthropology), we gain better understanding of the compromises and trade-offs that occurred during human evolution and history which contribute to our biological design today. The study of fossils, as well as the study of more recent human populations. in regard to their work distribution, demography, physical load, diet and nutrition, health, and migration improves our understanding of human biological variation nowadays and might suggest the etiological factors for many of the present-day diseases. Since we have at Tel Aviv University an ever-growing anthropological collection that spans a large time scale, we can examine biological changes over time in populations that lived in a limited geographical region. Therefore, confounders related to environmental conditions become less significant, which helps to enhance our hypotheses and conclusions about the past lives of these ancient populations.

### *What's the best thing about your job? What would you change if you could?*

I am fortunate to feel as though my work and my hobby are the same. There isn't a dull moment and I get to choose which research to carry out or to be involved in. I believe that I would be happy to add more field days although I already spend about a third of my time in excavations.

*Hila and her daughter, Emma, excavating at Manot Cave, located in the Galilee, Israel.*

***References:***

Borgel, S., Latimer, B., McDermott, Y., Sarig, R., Pokhojaev, A., Abulafia, T., Goder-Goldberger, M., Barzilai, O., and May, H. (2019). Early Upper Paleolithic Human Foot bones from Manot Cave, Israel. *Journal of Human Evolution* 102668, DOI: 10.1016/j.jhevol.2019.102668

Floreanova, K., Gilat, E., Koren, I. and May, H. (2020). Ear infection prevalence in prehistoric and historic populations of the southern Levant: A new diagnostic method. *International Journal of Osteoarchaeology*, DOI: 10.1002/oa.2873

Harney, É., May, H., Shalem, D., Rohland, N., Mallick, S., Lazaridis, I., Sarig, R., Stewardson, K., Nordenfelt, S., Patterson, N., Hershkovitz, I. and Reich, D. (2018). Ancient DNA from Chalcolithic Israel reveals the role of population mixture in cultural transformation. *Nature Communication* 9(1): 3336, DOI: 10.1038/s41467-018-05649-9

Hershkovitz, I., Marder, O., Ayalon, A., Bar-Matthews, M., Yasur, G., Boaretto, E., Caracuta, V., Alex, B., Frumkin, A., Goder-Goldberger, M., Gunz, P., Holloway, R. L., Latimer, B., Lavi, R., Matthews, A., Slon, V., Mayer, D. B., Berna, F., Bar-Oz, G., Yeshurun, R., May, H., Hans, M. G., Weber, G. W. and Barzilai, O. (2015). Levantine cranium from Manot Cave (Israel) foreshadows the first European modern humans. *Nature* 520(7546), 216-9, DOI: 10.1038/nature14134.

Hershkovitz, I., Weber, G. W., Quam, R., Duval, M., Grün, R., Kinsley, L., Ayalon, A., Bar-Matthews, M., Valladas, H., Mercier, N., Arsuaga, J. L., Martinón-Torres, M., Bermúdez de Castro, J. M., Fornai, C., Martín-Francés, L., Sarig, R., May, H., Krenn, V. A., Slon, V., Rodríguez, L., García, R., Lorenzo, C., Carretero, J. M., Frumkin, A., Shahack-Gross, R., Bar-Yosef Mayer, D. E., Cui, Y., Wu, X., Peled, N., Groman-Yaroslavski, I., Weissbrod, L., Yeshhurun, R., Tsatskin, A., Zaidner, Y. and Weinstein-Evron, M. (2018). The earliest modern humans outside Africa. *Science* 359 (6374): 456-459, DOI: 10.1126/science.aap8369

May, H. and Ruff, C. (2016). Physical burden and lower limb bone structure at the origin of agriculture in the Levant. *American Journal of Physical Anthropology* 161:26-36, DOI: 10.1002/ajpa.23003.

May, H., Sella-Tunis, T., Pokhojaev, A., Peled, N. and Sarig, R. (2018). Changes in mandible characteristics during the terminal Pleistocene to Holocene Levant and their association with dietary habits. *Journal of Archaeological Science: Reports*, 22: 413-419, DOI: 10.1016/j.jasrep.2018.03.020

Peleg, S., Pelleg Kallevag, R., Dar, G., Steinberg, N., Masharawi, Y. and May, H. (2020). New methods for sex and ancestry estimation using sternum and rib morphology. *International Journal of Legal Medicine*, 1-12, DOI: 10.1007/s00414-020-02266-4

Pokhojaev, A., Avni, H., Sella-Tunis, T., Sarig, R. and May, H. (2019). Changes in human mandibular shape during the Terminal Pleistocene-Holocene Levant. *Scientific Report* 9(1): 8799, DOI: 10.1038/s41598-019-45279-9

Sella-Tunis, T., Sarig, R., Cohen, H., Medlej, B., Peled, N. and May, H. (2017). Sex estimation using computed tomography of the mandible. *International Journal of Legal Medicine* 131(6): 1691-1700, DOI: 10.1007/s00414-017-1554-1

Sella-Tunis, T., Pokhojaev, A., Sarig, R., O'Higgins, P. and May, H. (2018). Human mandibular shape is associated with masticatory muscle force. *Scientific Reports* 8 (1): 6042, DOI: 10.1038/s41598-018-24293-3

## Dr Simon Underdown[20]

Dr Simon Underdown is a Reader in Biological Anthropology at Oxford Brookes University. His research primarily focuses on the co-evolution of humans and disease, specifically how human evolutionary processes can be reconstructed through identifying patterns in past human-diseases. He's undertaken fieldwork across the world, including South America, the Middle East and sub-Saharan Africa. He is a passionate science educator, holding the position of Chair of the Royal Anthropological Institute's Education Committee and is a Chartered Science Teacher, alongside appearing on radio, TV and in newspapers to discuss human evolution. Simon is former Chair (and current committee member) of the Society for the Study of Human Biology, and a member of the QAA Anthropology subject bench-marking panel.

### *What are your research interests and particular area of expertise?*

I'm something of an academic magpie really – interested in everything. Broadly my research focusses on the co-evolution of humans and disease and how patterns of human-disease interaction in the past can be used to reconstruct human evolutionary patterns and processes. I'm especially interested in the role played by diseases in shaping the adaptive environment during human evolution and the impact of disease exchange during contact between hominin species. If I was being more philosophical I suppose my research tackles questions about how humans in the past have responded to challenges presented by their environment in its widest sense. The evolution of the human mind has provided us with a unique (at least in extant hominins) ability to adaptationally grapple with selective pressures on a cognitive as well as biological level and only through designing collaborative inter-disciplinary research can we hope to understand how human evolution works. I take a holistic approach, combining information from fossils, artefacts, and ancient and modern DNA to attempt to reconstruct hominin behaviour and its underlying processes. But if I was being pithy and thinking in terms of one-liners then my research explores the intersection between human biology and cultural adaptation.

### *What originally drew you towards biological anthropology?*

Dinosaurs, and a certain celluloid American archaeologist with a fedora, leather jacket and a questionable understanding of excavation and international law. My first degree was in archaeology and it was that which stoked my interested in human evolution. In our first lecture we were told to put down our pens and just listen, then spend the next week reading Analytical Archaeology by David Clarke; both left a lasting impression on how I think about research and, indeed, how I approach teaching (a great lecture will always trump a million powerpoint slides). Bio anth is such a brilliant subject because there are no limits to the questions we ask. It's allowed me to work with great colleagues and carry out fieldwork across the world.

[20] Department of Anthropology, Oxford Brookes University; sunderdown@brookes.ac.uk

### *What was your PhD topic? How did you find your PhD experience?*

I did my PhD at Cambridge on Neanderthals with Rob Foley. I had a great time in a really exciting department. It's always joked that the whole world seems to pass through Cambridge but it's really true. Seminars from almost every leading figure in the field became normal very quickly. I learnt a lot from Rob, but above all the importance of not being constrained by a narrow definition of a subject or methodology. Another Cambridge alum, Roger Bacon, said much the same thing in 1620 – just because something has always been done in a particular way is not a good reason to keep doing so unquestionably.

### *After your PhD, what positions have you held and where?*

In the current job market, I'll whisper this... I was fortunate enough to get a lectureship at Oxford Brookes right out of my PhD. I'm still there, now Reader in Biological Anthropology and Director of the Research Centre for Environment and Society. I'm also a visiting fellow at the Center for Microbial Ecology and Genomics at the University of Pretoria.

*Simon at Wadi Faynan, Jordan.*

### *What current projects are you working on? Where do you hope these go in the future?*

My research projects at the moment are mostly focussed on sub-Saharan Africa and Southern Arabia. Obviously Covid-19 has had a massive impact on fieldwork and we're still trying to adapt and adjust. We've been working on ancient sedimentary DNA from sites across Southern Africa and Arabia (including Jebel Faya) and have just secured major funding for a three-year project in Oman. With my good friend Riaan Rifkin I'm also working on a project in Namibia exploring the Kalahari San hunter-gatherer skin and intestinal microbiome composition. Busy but fascinating work!

### *What project or publication or discovery are you most proud of?*

I think the best 'thing' I've ever found has nothing to do with human evolution. I was visiting a friend's project in Nazareth many years ago and found a beautiful pair of Crusader era column bases that had been flipped over to create a crude step. My favourite publication is a 2016 paper I co-wrote with my good friend and colleague Charlotte Houldcroft (Houldcroft and Underdown, 2016). Not only was it great fun to write and attracted a pleasing amount of press coverage, the ideas we developed about differential pathogen resistance and the impact of genetic exchange between closely related hominin species are being borne out almost weekly as new research reveals the impact of Neanderthal genes on the *Homo sapiens* genome – not bad for an idea we sketched out over a coffee.

*Simon at Pomongwe Cave in Zimbabwe.*

### *What do you think is the most revolutionary discovery in human evolution research over the last 5 years?*

There are too many to choose from! Human Evolution is a subject that benefits (and suffers) from a single find being able to radically change how we interpret the past. Obviously, this makes it really exciting to research but does mean frequent lecture updates! The analysis of ancient biomolecules has had a huge impact on how we think about human evolution as a process and the wide range of new 'species' discovered over the last 20 years has similarly knocked long-held theories on the head. But I think the most revolutionary change has not come from bones or stones but rather the way in which technology has transformed how we study human evolution. I'm old enough to remember fieldwork before wifi and mobile phones and while I do sometimes miss the experience of remoteness we had in those days, the ability to be able to instantly discuss finds with colleagues from a site and share data (and of course being able to keep in touch with family) is transformative. Likewise, the boon in sharing 3D scans of fossil material instantly has changed the rules of the game. 'Ownership' of a fossil is quite rightly becoming a thing of the past.

### *What is the best thing about your job and what is one thing you would change if you could?*

Working with amazing colleagues from a huge range of disciplines and getting to explore huge swathes of the world. If I had to pick the very best thing about being an academic, I'd have to say complete freedom to explore things that interest me (as long as I can bring in the money!). Being an academic is very odd in the 21st Century. I'm extremely lucky to be able to spend my time researching things that fascinate me and then tell people all about them. But it would be naive to think that the groves of academe do not have problems. If I had a magic wand, I would use it to provide a clear route towards long-term financial security for the hundreds if not thousands of doctoral, post-doctoral and associate lecturers in human evolution who are competing for jobs in an increasingly hostile environment. As a subject we are brilliant at attracting interest from the public (barely a week goes by without a human evolution story being in the news), but we do struggle to translate this into more full-time permanent jobs.

### *References:*

Houldcroft, C. J. and Underdown, S. J. (2016). Neanderthal genomics suggests a Pleistocene time frame for the first epidemiologic transition. *American Journal of Physical Anthropology* 160(3), 379-388. DOI: 10.1002/ajpa.22985

## Professor Lluis Quintana-Murci[21]

Professor Lluis Quintana-Murci is a population geneticist at the Pasteur Institute and Collège de France! He has led the Human Evolutionary Genetics Unit at the Pasteur Institute since 2007, and also currently holds the position of Professor of Human Genomics and Evolution at the Collège de France in Paris. Lluis is internationally renowned for his research on the genetic architecture of human populations and the role of genetic diversity in human adaptation. His team is particularly interested in how genomic data can be used to infer the past demographic history of our species, to explore how natural selection influences human diversity and to understand how pathogens have shaped human evolution.

### *What are your research interests and your particular area of expertise?*

My area of expertise is human population genetics, with an emphasis on understanding both the demographic past of human populations and the various ways through which they have adapted to local environments – climatic, nutritional or pathogenic – all along their journey around the world. I am particularly interested in two areas of the world: Central Africa, which hosts the largest group of living hunter-gatherers (the so-called rainforest hunter-gatherers or 'pygmies') and, more recently, the south Pacific.

I also have a strong interest in how pathogens and infectious diseases have shaped human evolution, since infectious agents have accompanied humans since their emergence in Africa. Understanding how natural selection imposed by pathogens has affected the diversity of our genomes is an alternative way to identify genes and biological functions that have play a key role in human survival against deadly infectious diseases, which highlights the value of dissecting the most natural experiment ever done: that of Nature (Quintana-Murci *et al.*, 2019).

In the context of selection and adaptation, I have a strong interest in how admixture, be it ancient or modern, has influenced human adaptation. In other words, when one faces a new environment to which they need to adapt, instead of waiting for an advantageous mutation to appear and increase in frequency in the population, why not rather admix with another population that already harbours the advantageous mutation in their gene pool? (Patin *et al.*, 2017) We have been quite focused on this topic over the last few years, in particular, on how ancient early Europeans acquired, through admixture with Neanderthals, advantageous variants involved in resistance to infectious diseases (Quach *et al.*, 2016).

### *What originally drew you towards population genetics?*

A combination of many things. I grew up in Mallorca, an island of the Mediterranean, so my contact with nature was quite important. I liked the sea a lot and my mother used to show me documentaries about Jacques Cousteau and his sea explorations (that was the only way I shut up immediately...since I was a very, very chatty child). These things conditioned me to like nature and I wanted to be a marine biologist. At the same time, ever since I was a child, I have always been very attracted by other cultures

[21] Human Evolutionary Genetics Unit, Institut Pasteur and Collège de France; lluis.quintana-murci@pasteur.fr

and people speaking other languages. I remember listening carefully on the beach to the languages spoken by tourists. I did love that; it was my own way of traveling. I've felt attracted to human diversity forever. All this, together with my strong interest in history, naturally brought me to population and evolutionary genetics, where diversity, history and biology form the basis of it.

***What was your PhD topic? Where did you complete your PhD and who was your supervisor?***

I did my PhD in the University of Pavia (Italy) under the direction of one of the first students of Luca Cavalli-Sforza: Silvana Santachiara-Benerecetti, who passed away a few months ago. Indeed, during the early 90's, when I was doing my PhD, Cavalli-Sforza spent a few months in the department, swinging between Pavia and Stanford. Pavia was a great place to do a PhD in population genetics, given the strong evolutionary and population genetics flavour of the whole department. Life-wise, it was less fun...having lived in Mallorca and Barcelona in the 'post-movida' 80's period, living in a small provincial Italian city was like travelling back to the 50's during my grandmother's childhood! The topic of my PhD was the evolutionary history of eastern African populations through the study of mitochondrial DNA and Y chromosome markers, which was the best you could do in human population genetics in the early 90's.

*Lluis with Luca Cavalli-Sforza (the father of modern human population genetics) during his PhD in Italy (around 1998). Photo credit: Franz Manni*

### *What were the findings from your PhD?*

They were super cool. We first found a mitochondrial DNA marker of Indian origin in various Ethiopian populations, and we wondered whether it was the result of recent admixture or if it was an old lineage present in eastern Africa that was brought to India. By enlarging our sample sizes from eastern Africa and several Asian and European populations, we showed that indeed we had discovered an old marker that supported a coastal route out of Africa of modern humans 60,000 years ago, starting from the horn of Africa and following the coasts of south Asia. This was the first genetic evidence of a coastal route of exit of modern humans out of Africa and was published in *Nature Genetics* in 1999 (Quintana-Murci *et al.*, 1990).

### *After your PhD, where have you worked? Where has been your favourite place to work?*

After my PhD, I have been essentially living in Paris for the whole time. In theory, I came here 20 years ago for a post-doc and I never left! Workwise, the Pasteur Institute is great, and Paris....is Paris! I immediately felt in love with this city. It is a good balance between a southern city and a northern city. Though don't think that I am chauvinist, since I am not French! (well, now I am also French). But seriously, I really like Paris both to work and to live. Having said that, I have spent periods in other cities as an invited researcher or professor. I have great memories of Tucson (Arizona) where I spent some months in Mike Hammer's lab. I loved that city, a bit calm for my taste, but the nature was amazing. I also spent a summer at Rockefeller University two years ago...one of the best experiences of my life. I do love NYC!

### *What current projects are you working on at the Pasteur Institute Where do you hope these go in the future?*

We continue to work on the demographic and adaptive history of humans, but with a stronger focus now on the South Pacific. This region of the world is a land of contrasts, as it was first peopled just after the out-of-Africa exodus, around 45,000 years ago, then no migrations occurred until around 5,000 years ago when Austronesian-speaking peoples, most likely from Taiwan, entered the region. It is also super interesting that some populations from the Pacific present the highest worldwide levels of combined Neanderthal and Denisovan ancestry. We have dissected the demographic past of this region and explored how the 'archaic' ancestry found in these populations participated in their adaptation to the island environments of the Pacific (Choin *et al.,* 2021).

On the other hand, we maintain a strong interest in immunity and infectious diseases. In particular, we are exploring the genetic, epigenetic and environmental factors that drive our differences in immune responses (Piasecka *et al.,* 2018). We are all different, we are all diverse (fortunately!), and we want to understand the sources of such diversity. We are also studying how ethnic background, age, sex and metabolism influence our susceptibility to infection, using as a model influenza A virus and...SARS-CoV-2!

### *What advice would you give to a student interested in your field of research?*

Just one. Be passionate. If you like science, all else will come along perfectly. Because if you like science, you will naturally work hard at it. Some people really insist that we should maintain a good work-life balance. And this is very true. But sometimes I don't really understand it, as I do many things in my life, science, writing, reading, gardening, watching movies, eating, etc...but when I do science, I don't think 'Now, time to work'. For me, science is totally integrated into my 'life'. Hopefully that makes sense.... I guess what I mean is that the only healthy advice I can give is: if you like science, go for it, you will find a job and you will enjoy it. A bit of mobility also helps to find a job.

*Lluis in Polynesia (Moorea, 2017), an area of the world where much of his research is now focussed.*

**What do you think has been the most revolutionary discovery in evolutionary genetics over the last 5 years?**

I am obviously biased but, for me, it has been the discovery, bolstered by the progress in paleogenomics, that admixture with ancient hominins like Neanderthals or Denisovans helped early modern humans to adapt to their local environments, a phenomenon known as 'adaptive introgression'. As some of my colleagues say, 'this part of Neanderthal that is within us' helped early Eurasians to adapt to cold climates and pathogens, viruses in particular, among other phenotypes.

**If you were not an evolutionary geneticist, what would you be?**

That's tricky. So many things. If I had to remain in science, I would be a marine biologist, my first passion, and work on the behavioural biology of Cetaceans. I am fascinated by the complexity of their social relationships as well as by their mode of communication. Outside of science, I would have liked to be a writer, which actually I do a lot in science anyway, and it is also something that I love. I like the mixed feeling of suffering and satisfaction that comes at the same time when writing. Yes, I would have liked to be a writer.

***References:***


Choin, J., Mendoza-Revilla, J., Arauna, L. R., Cuadros-Espinoza, S., Cassar, O., Larena, M., Min-Shan, K. A., Harmant, C., Laurent, R., Verdu, P., Laval, G., Boland, A., Olaso, R., Deleuze, J. F., Valentin, F., Ko, Y. C., Jakobsson, M., Gessain, A., Excoffier, L., Stoneking, M., Patin, E. and Quintana-Murci, L. (2021). Genomic insights into population history and biological adaptation in Oceania. Nature (in press).

Patin, E., Lopez, M., Grollemund, R., Verdu, P., Harmant, C., Quach, H., Laval, G., Perry, G. H., Barreiro, L. B., Froment, A., Heyer, E., Massougbodji, A., Fortes-Lima, C., Migot-Nabias, F., Bellis, G., Dugoujon, J. M., Pereira, J. B., Fernandes, V., Pereira, L., van der Veen, L., Mouguiama-Daouda, P., Bustamante, C. D., Hombert, J. M. and Quintana-Murci, L. (2017). Dispersals and genetic adaptation of Bantu-speaking populations in Africa and North America. *Science* 356(6337):543-546, DOI: 10.1126/science.aal1988

Piasecka, B., Duffy, D., Urrutia, A., Quach, H., Patin, E., Posseme, C., Bergstedt, J., Charbit, B., Rouilly, V., MacPherson, C. R., Hasan, M., Albaud, B., Gentien, D., Fellay, J., Albert, M. L., Quintana-Murci, L. and Milieu Intérieur Consortium (2018). Distinctive roles of age, sex, and genetics in shaping transcriptional variation of human immune responses to microbial challenges. *Proceedings of the National Academy of Science of the United States of America.* 115(3):E488-E497, DOI: 10.1073/pnas.1714765115

Quach, H., Rotival, M., Pothlichet, J., Loh, Y. E., Dannemann, M., Zidane, N., Laval, G., Patin, E., Harmant, C., Lopez, M., Deschamps, M., Naffakh, N., Duffy, D., Coen, A., Leroux-Roels, G., Clément, F., Boland, A., Deleuze, J. F., Kelso, J., Albert, M. L. and Quintana-Murci, L. (2016). Genetic Adaptation and Neandertal Admixture Shaped the Immune System of Human Populations. *Cell* 167(3):643-656.e17, DOI: 10.1016/j.cell.2016.09.024

Quintana-Murci, L. (2019). Human Immunology through the Lens of Evolutionary Genetics. Cell. 177(1):184-199, DOI: 10.1016/j.cell.2019.02.033

Quintana-Murci, L., Semino, O., Bandelt, H.J., Passarino, G., McElreavey, K. and Santachiara-Benerecetti, A.S. (1999). Genetic evidence of an early exit of Homo sapiens from Africa through eastern Africa. *Nature Genetics* 23, 437–441, DOI: 10.1038/70550.